Modernizing Oracle Tuxedo Applications with Python

A practical guide to using Oracle Tuxedo in the 21st century

Aivars Kalvāns

BIRMINGHAM—MUMBAI

Modernizing Oracle Tuxedo Applications with Python

Copyright © 2021 Packt Publishing

Group Product Manager: Aaron Lazar

Publishing Product Manager: Denim Pinto

Senior Editor: Rohit Singh

Content Development Editor: Tiksha Lad

Technical Editor: Rashmi Subhash Choudhari

Copy Editor: Safis Editing

Project Coordinator: Deeksha Thakkar

Proofreader: Safis Editing

Indexer: Rekha Nair

Production Designer: Prashant Ghare

First published: March 2021

Production reference: 1190321

Published by Packt Publishing Ltd.

Livery Place

35 Livery Street

Birmingham

B3 2PB, UK.

ISBN 978-1-80107-058-4

www.packt.com

Contributors

About the author

Aivars Kalvans is a developer, software architect, and consultant. He has been using Oracle Tuxedo (formerly BEA Tuxedo) since version 8 in 2003 to develop the Tieto Card Suite payment card system using the C++, Python, and Java programming languages. During his almost 19-year career at Tieto, Aivars has been involved in many projects related to payment card issuing, acquiring, and utility payments through mobile phones, ATMs, and POS terminals.

I want to thank my lovely wife, Anete, and sons, Kārlis, Gustavs, and Leo,
for making life much more interesting.

About the reviewer

Klāvs Dišlers is a software developer at Moon Inc. He has over 8 years of experience in maintaining and improving Oracle Tuxedo and Python-based parts of the Tieto Card Suite payment system. During his career, he has mainly focused on developing libraries and automated testing tools to improve the day-to-day tasks of his coworkers and clients.

I would like to thank my wonderful wife, Kristīne; my precious daughter, Marija; and my curious son, Valters, for making my life meaningful.

Table of Contents

Section 2:
The Good Bits

5

Developing Servers and Clients

6

Administering the Application Using MIBs

7

Distributed Transactions

8

Using Tuxedo Message Queue

9
Working with Oracle Database

Section 3: Integrations

10
Accessing the Tuxedo Application

11
Consuming External Services in Tuxedo

12
Modernizing the Tuxedo Applications

Assessments

Other Books You May Enjoy

Index

Preface

I started working with BEA Tuxedo version 8 in 2003 by writing code in the C programming language. Tuxedo seemed like a magical framework that just worked and enabled several teams to develop microservices (we called them *components* then) and integrate them into a single distributed system in the end. Despite being developed in the 1980s, Oracle Tuxedo still runs a significant portion of critical infrastructure and is not going away any time soon. Yet the API provided by Oracle Tuxedo is still based on the same C programming language. Developing business logic in the C programming language is overkill in 2021; we should be using better tools for that. For developer productivity, nothing beats Python.

After 17 years, I have developed my own mental model of Tuxedo by writing and debugging code every day. On one hand, it is on a level of abstraction higher than the Tuxedo API and I operate in terms of C++ and Python wrappers for the lower-level Tuxedo API. On the other hand, I dive below Tuxedo abstractions into UNIX operating system processes, message queues, and other UNIX APIs. For me, that explains the application behavior and quirks I have observed. I have also used the same approach when teaching new recruits about Tuxedo.

So, this book combines what I wish I knew when I started developing with Tuxedo and the programming language and library I wish I had in my toolbox. I hope it gives you a different viewpoint than the Tuxedo documentation does and makes your Tuxedo journey more pleasant.

Who this book is for

It will be the greatest help for new developers learning to maintain existing Tuxedo applications and trying to integrate them with other parts of the infrastructure. QA engineers will learn an easier way to automate Tuxedo application tests. I believe experienced Tuxedo developers will find out about new Tuxedo features that help to implement features unimaginable before.

What this book covers

Chapter 1, *Introduction and Installing Tuxedo*, covers the history of Tuxedo, some core concepts, and the important aspects of it. We will also prepare a development environment by installing Tuxedo in non-interactive mode.

Chapter 2, *Building Our First Tuxedo Application*, will teach us how to develop a simple to-upper application using Python just like the documentation and examples do for the C programming language. We will examine the running application processes with both Tuxedo and UNIX tools.

Chapter 3, *Tuxedo in Detail*, introduces the Bulletin Board, clients, and servers. We will learn how queues are used to communicate between clients and servers and how the service abstraction is used for load balancing.

Chapter 4, *Understanding Typed Buffers*, is dedicated to the message formats that are exchanged between clients and servers.

Chapter 5, *Developing Servers and Clients*, will teach us how Tuxedo servers expose services and how clients can access them.

Chapter 6, *Administering the Application Using MIBs*, covers the programmatic configuration of the application and upgrades with zero downtime. We will learn how to extract statistics for monitoring that existing tools do not provide.

Chapter 7, *Distributed Transactions*, will – better late than never – teach us about one of Tuxedo's main features: transactions. We will experiment with timeouts so you do not have to do it on the production system.

Chapter 8, *Using Tuxedo Message Queue*, will cover using the Tuxedo /Q component for improving the reliability of the application.

Chapter 9, *Working with Oracle Database*, covers all the necessary steps to use Oracle Database from the Tuxedo application and execute SQL in local and global transactions.

Chapter 10, *Accessing the Tuxedo Application*, will show us how to access a remote Tuxedo application over the network or expose friendly web services from the Tuxedo application itself.

Chapter 11, *Consuming External Services in Tuxedo*, will cover accessing external services from the Tuxedo application in a way that fits best with other parts of the application.

Chapter 12, *Modernizing the Tuxedo Application*, introduces the NATS messaging system and develops an example of the Tuxedo application and the NATS application calling services provided by the other application.

To get the most out of this book

You will need a computer running either Linux and Docker or Microsoft Windows and Docker Desktop. You will need an Oracle account for downloading the Tuxedo version 12.2.2 installation. For *Chapter 9, Working with Oracle Database*, you will need an Oracle Database instance. You should also have at least basic Python and Linux knowledge for editing files and running commands. All the code examples were tested on Microsoft Windows 10 running Docker Desktop 2.5, Oracle Linux 8.2 with Python 3.6.8, and Tuxedo 12.2.2.

If you are using the digital version of this book, we advise you to type the code yourself or access the code via the GitHub repository (link available in the next section). Doing so will help you avoid any potential errors related to the copying and pasting of code.

Download the example code files

You can download the example code files for this book from GitHub at `https://github.com/PacktPublishing/Modernizing-Oracle-Tuxedo-Applications-with-Python`. In case there's an update to the code, it will be updated on the existing GitHub repository.

We also have other code bundles from our rich catalog of books and videos available at `https://github.com/PacktPublishing/`. Check them out!

Code in Action

Code in Action videos for this book can be viewed at `http://bit.ly/3csxr9N`.

Download the color images

We also provide a PDF file that has color images of the screenshots/diagrams used in this book. You can download it here:

`https://static.packt-cdn.com/downloads/9781801070584_ColorImages.pdf`

Conventions used

There are a number of text conventions used throughout this book.

`Code in text`: Indicates code words in text, database table names, folder names, filenames, file extensions, pathnames, dummy URLs, user input, and Twitter handles. Here is an example: "Mount the downloaded `WebStorm-10*.dmg` disk image file as another disk in your system."

A block of code is set as follows:

```
import tuxedo as t
_, _, res = t.tpcall(
    ".TMIB",
    {
        "TA_CLASS": "T_DOMAIN",
        "TA_OPERATION": "GET",
    },
)
```

When we wish to draw your attention to a particular part of a code block, the relevant lines or items are set in bold:

```
GROUP1 LMID=tuxapp GRPNO=1 TMSNAME=TMS TMSCOUNT=2
*SERVERS
"ping.py" SRVGRP=GROUP1 SRVID=1
    REPLYQ=Y MAXGEN=2 RESTART=Y GRACE=0
    MIN=1 MAX=1
"api.py" SRVGRP=GROUP1 SRVID=20
    REPLYQ=Y MAXGEN=2 RESTART=Y GRACE=0
    MIN=1 MAX=1
```

Any command-line input or output is written as follows:

```
tmloadcf -y ubbconfig
echo crdl -z `pwd`/tlog -b 200 | tmadmin
echo crlog -m tuxapp | tmadmin
```

> **Tips or important notes**
> Appear like this.

Get in touch

Feedback from our readers is always welcome.

General feedback: If you have questions about any aspect of this book, mention the book title in the subject of your message and email us at customercare@packtpub.com.

Errata: Although we have taken every care to ensure the accuracy of our content, mistakes do happen. If you have found a mistake in this book, we would be grateful if you would report this to us. Please visit www.packtpub.com/support/errata, selecting your book, clicking on the Errata Submission Form link, and entering the details.

Piracy: If you come across any illegal copies of our works in any form on the Internet, we would be grateful if you would provide us with the location address or website name. Please contact us at copyright@packt.com with a link to the material.

If you are interested in becoming an author: If there is a topic that you have expertise in and you are interested in either writing or contributing to a book, please visit authors.packtpub.com.

Reviews

Please leave a review. Once you have read and used this book, why not leave a review on the site that you purchased it from? Potential readers can then see and use your unbiased opinion to make purchase decisions, we at Packt can understand what you think about our products, and our authors can see your feedback on their book. Thank you!

For more information about Packt, please visit packt.com.

Section 1:
The Basics

In this section, we will introduce Tuxedo and get a development environment ready for the rest of the book. We will cover the Tuxedo client-server architecture and messages sent between clients and servers in depth from both the Tuxedo and UNIX sides.

This section has the following chapters:

- *Chapter 1, Introduction and Installing Tuxedo*
- *Chapter 2, Building Our First Tuxedo Application*
- *Chapter 3, Tuxedo in Detail*
- *Chapter 4, Understanding Typed Buffers*

1
Introduction and Installing Tuxedo

In this chapter, we will introduce Tuxedo and learn why it is still relevant and in use even today. I will try to convince you that the Python programming language is a good choice when it comes to writing, extending, and improving Tuxedo applications. And finally, we will create a development environment for learning Tuxedo development using Python.

We will cover the following topics in this chapter:

- Introducing Tuxedo
- Understanding Tuxedo through modern lens
- Reviewing the need for Python
- Installing Tuxedo and Python

By the end of this chapter, you will have a good understanding of Tuxedo and have your environment set up for building your own applications.

Technical requirements

To follow the instructions in this chapter, you will require the following:

- Docker for Linux, or Docker Desktop for Windows 10

- A 64-bit processor

- At least 3 GB of free disk space

- At least 4 GB of RAM, but 8 GB is strongly recommended

You can find the code files for this chapter on GitHub at `https://github.com/PacktPublishing/Modernizing-Oracle-Tuxedo-Applications-with-Python/tree/main/Chapter01`. The Code in Action video for the chapter can be found at `https://bit.ly/3rWX4Gh`.

Introducing Tuxedo

Tuxedo started in the year 1983 at AT&T as a framework for building high-performance, distributed business applications. It has its roots in the DUX (Database for Unix) and TUX (Transactions for Unix) projects, which were combined into a client-server communication framework with support for transactions under the name TUXEDO. Tuxedo stands for **Transactions for Unix, Extended for Distributed Operations**.

Tuxedo evolved from a client-server framework of release 1.0 to support high availability in release 2.0, and then to distributed and heterogeneous system support in releases 3.0 and 4.0. By release 6.1 and the year 1995, Tuxedo had reached maturity. A book entitled *The TUXEDO System* was published and it described Tuxedo as we know it today. Tuxedo continued to improve and advance over the later years and gained additional features until the most recent release, 12.2.2, but the core of it is pretty close to what it was in 1995.

Some of the innovations made by the engineers of Tuxedo became standardized; most notably, the XA interface is still implemented by databases and used for distributed transactions even in modern Java applications. Less known is the XATMI standard describing the core Tuxedo API. You may be familiar with the XA and XATMI specifications from X/Open Company, but not many people know that the inspiration came from Tuxedo.

Another fun fact is that Tuxedo enabled **Service-Oriented Architecture (SOA)** before the term was coined, and it enabled us to write microservices and even transactional microservices before the microservice architectural style became popular. Of course, a framework that is more than three decades old does not fit modern ideas when it comes to application servers, middlewares, microservice frameworks, and so on, but if you take a step back, you will see similarities.

A Unix-inspired application server

Tuxedo is an application server for the C and COBOL programming languages. Application servers, as we know them today for other programming languages, are typically a single process with multiple threads. That leads to isolation problems when multiple applications are running in the same application server. Some languages ignore that, while others try to minimize it by using different class loaders and other techniques. But still, all applications use the same heap memory. A single misbehaving application will affect other applications either by consuming all CPU resources or consuming too much memory.

Tuxedo follows the original Unix philosophy of having many (small) processes that communicate with each other. Such a process is called a **server** and a Tuxedo application consists of many servers. Another kind of process is a **client**, which accesses facilities provided by the server but does not provide facilities itself. Unix processes give memory isolation out of the box for Tuxedo applications: memory corruption by one server will not affect others. Processes are also scheduled by the operating system to give each of them a fair share of CPU time. A process can be assigned a priority or be run under different user accounts.

Communication between clients and servers is done using Unix System V ("System Five") **inter-process communication** (**IPC**) mechanisms: message queues, shared memory segments, and semaphores. Distributed computing is added transparently on top of that by forwarding messages from queues over the TCP/IP network to a different machine. All Tuxedo applications are distributed applications by design: communication is done by passing messages; a receiver may fail before or after processing the message; the sender may fail before receiving the response. The possibility of failure is exposed in the API.

After getting an idea of what Tuxedo is, let's learn more about it in detail in the next section.

Understanding Tuxedo through modern lens

In this section, we will look at various aspects of Tuxedo through a modern lens. We cannot cover all of them, but I believe each sections covers an aspect that is still relevant today.

Availability

If a Tuxedo server process crashes, it will get restarted. Multiple copies of the same executable may be run at the same time for redundancy. Multiple machines may be used to improve availability when one of them crashes. Tuxedo will load balance requests between the processes and machines that are up and running to minimize response time.

All Tuxedo applications have an administrative process that looks for servers that are stuck or have crashed and restarts them.

It is also possible to reconfigure a Tuxedo application without downtime. Adding more servers, and changing and reconfiguring existing ones – it can all be done while the application is running. Also, new versions of the software can be deployed without interrupting business processes.

Performance

The Tuxedo framework itself is fast and requires few resources. The performance cost on top of your application code will be very low.

This is partially because of the IPC queues used for the communication mechanism. The roundtrip latency, that is, sending a request and receiving a response, is less than 40 microseconds. Yes, that is 0.000040 seconds for a service call between two processes on the same machine.

The other reason for such performance is at-most-once delivery semantics, also known as fire-and-forget. It provides no message delivery guarantees and the sender is responsible for retrying the request and coming up with a way for a receiver to filter duplicate requests. If that sounds scary at first, this approach is similar to the UDP protocol, which offers better performance compared to TCP/IP. The application can be made reliable by using reliable queues or other mechanisms.

Oracle Tuxedo, along with Oracle Database, was used in transaction processing the TPC-C benchmark to achieve more than 500,000 transactions/second (`http://www.tpc.org/tpcc/results/tpcc_results5.asp`). Hundreds and thousands of business transactions per second are processed by Tuxedo running on smaller servers and laptops. To give you some perspective, Bitcoin is limited to around 7 transactions per second.

Services

Tuxedo is a message-oriented middleware. It works by sending and receiving messages through IPC queues. However, a typical Tuxedo API does not mention IPC queues as other message-oriented middlewares do. Tuxedo has an abstraction on top of the queues called a **service**. Each Tuxedo server implements one or more services. A service may be backed by a single queue or by multiple queues, or multiple services may share the same queue. All of that is transparent to the application code. A service might even be running on a different machine, but that is a concern for the administrator configuring the application, not the developer.

What about microservices or macroservices? Tuxedo is neutral in this regard: you are free to implement one big monolith service or many smaller ones, each using a different storage functionality and implementation language. It is a design choice, not a technical decision. When you do implement your application as many smaller services, Tuxedo will help you to make them transactional if you want to.

Polyglot development

Tuxedo natively supports the C and COBOL programming languages and Tuxedo's development tools produce output in one of those languages.

Since you can write C-like code in C++, you can use C++ to develop Tuxedo applications as long as you take care of exception propagation and typecasting, and avoid some pitfalls. Tuxedo comes with support for writing Java code; however, your Java code will look more like C code than idiomatic and modern Java. Other languages, such as PHP, Python, and Ruby, are supported through a technology called **Service Component Architecture (SCA)**.

Transactions

Distributed transactions, or XA transactions, have a bad name these days, but I blame this on poor implementation in application servers. In Tuxedo, they work like a charm. You should design your application to avoid the need for distributed transactions, but when you need the consistency and are tempted to implement a solution with compensating transactions, just let Tuxedo do its work.

XATMI

X/Open ATMI (XATMI) stands for the **Application to Transaction Monitor Interface** and was based on existing Tuxedo API calls. It introduces **typed buffers**, which are used to exchange messages between parts of the application with the API to allocate, inspect, and release them. It also describes what a **service** is and how it is implemented. Finally, it specifies APIs for different messaging patterns: synchronous and asynchronous calls and conversational paradigms.

However, this standard captured what Tuxedo was capable of at some point in time. Tuxedo has since gained new APIs and supports typed buffers with more features and is better suited for complex applications. Think of XATMI as the SQL standard for databases: while the API is standardized, there are plenty of behavior differences not covered by the standard, and Tuxedo's XATMI will always be one of a kind.

Tuxedo today

Tuxedo became a product of the Oracle Corporation in 2008 and is now a part of Oracle Fusion Middleware, with the latest stable release being 12.2.2 in 2016. Tuxedo is a commercial, closed source product. With the current trend of relying on open source software, we will not see Tuxedo gaining huge popularity and acceptance. Most likely, it will not be used to develop new applications unless you use it already in your organization.

However, Tuxedo applications run a significant part of critical infrastructure in banking, telecommunications, payments, and many others. Just like COBOL, it is not going to disappear in the next few years; Tuxedo is here to stay for years to come. Those applications have to be maintained and enhanced and learning about Tuxedo might be a future job for you.

With detailed knowledge of Tuxedo, we are now ready to move on to the next section, where we will try to justify the use of the Python programming language.

Reviewing the need for Python

As I am writing this book, Python is the third most popular language according to the TIOBE Programming Community index. Given current trends, it may surpass Java and secure second place by the time this book is published. The only other language more popular than Python and Java is C.

So why should you choose Python instead of C or C++, which is supported natively by Tuxedo? Well, Python is simply a more productive tool for writing application logic. Unlike C, Python comes with "batteries included" and contains tools and libraries for many tasks. If the libraries included are not enough, Python has standard tools to download and install open source libraries for the missing functionality.

Python is a dynamic language and has some problems because of its dynamic nature, but, at the same time, it makes Python code easier to test by using mocks for database access and Tuxedo service calls. The same mocking enables you to migrate code away from Tuxedo if you choose to do so. It serves as an abstraction layer on top of Tuxedo, isolating your code from some of Tuxedo's APIs.

Even if the application is written in C or COBOL, using Python is beneficial for testing and quality assurance. It can be used for building a quick prototype before implementing it in C. There are plenty of good reasons to give it a try.

Tuxedo already comes with Python support, but sadly, the SCA standard did not gain popularity and is considered dead. Nothing prevents you from using SCA, but that is another API to learn in addition to XATMI and it exposes fewer features than Tuxedo provides. We will use something that does not hide the powerful XATMI and Tuxedo's improvements on top of it.

There are several open source libraries for developing Tuxedo applications using Python. This book will use Python's `tuxedo` module for all examples, but many examples can be implemented with slightly outdated `tuxmodule` or `tux_oracle` modules as well. And since those are open source modules, you can always add missing functionality yourself.

Installing Tuxedo and Python

Oracle provides Dockerfiles and commands to build Docker images for Oracle products on GitHub (`https://github.com/oracle/docker-images`). Tuxedo is no exception, although Oracle considers it *old content* and has moved into an archive location where it is available today. Therefore, we will create more up-to-date Docker images ourselves.

To install Tuxedo, you must first download it from the Oracle website (`http://www.oracle.com/technetwork/middleware/tuxedo/downloads/index.html`). You will have to create an account if you don't already have one. Tuxedo is available for different operating systems: AIX, Linux, HP-UX, Solaris, and Microsoft Windows. For all examples in this book, we will use the latest release of Oracle Tuxedo 12cR2 (12.2.2) and an installation package that includes all the required add-ons called "Oracle Tuxedo 12cR2 (12.2.2)" (including Tuxedo, SALT, and Jolt) for Linux x86-64 (64-bit). The downloaded file will be named `tuxedo122200_64_Linux_01_x86.zip`.

Create a new folder and put the downloaded `tuxedo122200_64_Linux_01_x86.zip` file there. You will need to create `tuxedo1222.rsp` and `Dockerfile` files in the same folder as the downloaded file.

We start with `tuxedo1222.rsp`, which contains answers for Oracle Tuxedo installer so that it can be installed in silent mode:

```
RESPONSEFILE_VERSION=2.2.1.0.0
FROM_LOCATION="/home/oracle/Disk1/stage/products.xml"
ORACLE_HOME="/home/oracle/tuxhome"
ORACLE_HOME_NAME="tuxhome"
INSTALL_TYPE="Custom Install"
DEPENDENCY_
LIST={"tuxedoServer:12.2.2.0.0","atmiClient:12.2.2.0.0"}
TLISTEN_PASSWORD="oracle"
```

This is a minimized response file that contains only the required responses. Here are the most important things:

- `FROM_LOCATION` gives the location of the unpacked ZIP file.
- `ORACLE_HOME` and `ORACLE_HOME_NAME` say that Tuxedo will be installed under `/home/oracle/tuxhome`.
- `INSTALL_TYPE` says we will install only selected components.
- `TOPLEVEL_COMPONENT` and `DEPENDENCY_LIST` say that only the *core Tuxedo* should be installed.

Then we need a `Dockerfile` file that contains instructions for building a Docker image:

```
FROM oraclelinux:8
RUN yum -y install oracle-release-el8 && \
    yum -y install \
            libnsl java-devel gcc-c++ python3-devel \
            unzip file hostname which sudo && \
    rm -rf /var/cache/yum
RUN groupadd oracle && \
        useradd -m -g oracle -s /bin/bash oracle && \
        echo 'oracle ALL=(ALL) NOPASSWD:ALL' \
        >> /etc/sudoers.d/oracle
USER oracle
```

```
COPY tuxedo1222.rsp tuxedo122200_64_Linux_01_x86.zip /home/
oracle/
ENV ORACLE_HOME=/home/oracle/tuxhome \
    JAVA_HOME=/etc/alternatives/java_sdk
RUN cd ~/ && \
      jar xf tuxedo122200_64_Linux_01_x86.zip && \
      cd ~/Disk1/install && \
      chmod -R +x * && \
      ./runInstaller.sh -responseFile ~/tuxedo1222.rsp \
          -silent -waitforcompletion && \
      rm -rf ~/Disk1 && \
      rm -f ~/tuxedo1222.rsp ~/tuxedo122200_64_Linux_01_x86.zip
ENV TUXDIR=/home/oracle/tuxhome/tuxedo12.2.2.0.0
ENV PATH=$PATH:$TUXDIR/bin
ENV LD_LIBRARY_PATH=$LD_LIBRARY_PATH:$TUXDIR/lib
USER root
RUN pip3 install tuxedo
USER oracle
WORKDIR /home/oracle
```

Once you have it, you can run the following command:

```
docker build -t tuxpy .
```

This command creates the Docker image. While it runs for several minutes, let's take a look at the most important steps:

1. The Docker image will be created from Oracle Linux 8. Other Linux distributions may also be used, but the choice of Oracle Linux will come in handy when we start using the Oracle database.

2. We run the yum package manager to install Python version 3, Java for running the Oracle Tuxedo installer, and the GCC compiler for building a Python module.

3. We create a regular Linux user named oracle and give permissions to run sudo for installing other software packages.

4. Once all the files are put into the container, the Tuxedo installation is unpacked.

5. The Tuxedo installation is run in non-interactive mode with a response file containing all the necessary inputs. Tuxedo can also be installed using a graphical interface or console interface like all other Oracle products, which is handy if you don't use containers.

6. We export the `TUXDIR` environment variable that points to the directory where Tuxedo is installed and set up program and library paths to include Tuxedo binaries.

7. After that is done, the Python `tuxedo` module is installed using the `pip3` tool.

After that, you can start the newly created image by using the following command:

```
docker run -ti --privileged tuxpy bash
```

If you are using Docker Desktop on Microsoft Windows, you may need to add `winpty` in front of the command:

```
winpty docker run -ti --privileged tuxpy bash
```

The `--privileged` parameter gives extended privileges to the container that will be needed in *Chapter 3, Tuxedo in Detail* to resize message queues.

If you have any preferences for a text editor or other tools, you can install them by using `sudo yum install` and the package name. As an exercise, take a look at the Python examples that use SCA under `$TUXDIR/samples/sca/simp_python`. It will make you appreciate the simplicity of your first Tuxedo application in Python, which we will create in the next chapter.

Summary

In this chapter, we introduced the decades-old Tuxedo framework for building distributed applications and technology that is still relevant today. We also looked at the current state of Tuxedo through the eyes of a fellow Tuxedo user and why the Python programming language is a good choice for development in the year 2021 and beyond. Lastly, we prepared a development environment by installing Tuxedo and Python in a Docker image. While it might not be a good idea to use a Docker container as the main development machine, it makes a nice throwaway sandbox for trying out code samples.

These topics helped you to acquire basic knowledge of Tuxedo and set up a working environment for creating applications. We will move things on in the next chapter, where you will build your first application.

Questions

1.   Which programming languages are supported by Tuxedo?

2.   How old is Tuxedo?

3.   What kind of queues does Tuxedo use?

Further reading

- *Getting Started with Oracle Tuxedo*, by Biru Chattopadhayay, describes Tuxedo from a different angle and provides the perspective of a Tuxedo application administrator.

2
Building Your First Tuxedo Application

Just like the first program in most programming languages is *Hello, world!*, any developer training for Tuxedo starts by implementing *to upper* service, converting a string into uppercase letters. There is a `simpapp` (`$TUXDIR/samples/atmi/simpapp`) application in Tuxedo code samples written in C if you are curious. So, let's do the same in Python!

In this chapter, we will cover the following topics:

- Creating a server
- Creating a client
- Creating a configuration file for Tuxedo
- Starting the application
- Administering the application
- Stopping the application

By the end of this chapter, you will have a good understanding of all the aspects that make up the Tuxedo application and will know how to perform the most basic administration tasks.

Technical requirements

All the code for this chapter is available at `https://github.com/PacktPublishing/Modernizing-Oracle-Tuxedo-Applications-with-Python/tree/main/Chapter02`.

The Code in Action video for the chapter can be found at `https://bit.ly/3ty5ZOy`.

Creating a server

Let's create our `toupper.py` server with the following content:

```
#!/usr/bin/env python3
import sys
import tuxedo as t
class Server:
    def TOUPPER(self, req):
        return t.tpreturn(t.TPSUCCESS, 0, req.upper())
t.run(Server(), sys.argv)
```

The implementation of the Tuxedo server in Python starts with a shebang line telling you to use the Python 3 interpreter to run the file. And while we're at it, the file must be executable:

```
chmod +x toupper.py
```

We import the `sys` module to access command-line options and the `tuxedo` module for using Tuxedo. For the book, we will use a shorter single-letter name, `t`, for the `tuxedo` module so that the code fits on one page without breaking the lines, but you should not sacrifice readability in your production code.

Tuxedo servers must be implemented as a class and the method name must match the desired service name. In this case, the service will be called `TOUPPER`. All service calls receive a single argument, request data, which is referred to as a **typed buffer**.

Then, the service ends by returning the following three values as the result, using the `tpreturn` function:

- Success (`TPSUCCESS`) or failure (`TPFAIL`), which mostly indicates to Tuxedo whether the transaction should commit or roll back
- Return code (`0`) to indicate the application-level result
- Response data, which is request data converted to uppercase

After we have defined the server class, we run the Tuxedo server by giving the instance of the server class implementing the functionality (`Server()`) and command-line options (`sys.argv`). Command-line parameters are important to Tuxedo's runtime for knowing how to join the application.

With our server ready, our next step is to create a client.

Creating a client

Create a `client.py` file with the following content:

```
import tuxedo as t
_, _, res = t.tpcall("TOUPPER", "Hello, world!")
print(res)
```

The Tuxedo client code is even simpler: just import the `tuxedo` module.

Then, use the synchronous remote procedure (service) call with the `tpcall` function and specify that we are calling the `TOUPPER` service and the string we want to convert to uppercase.

As a result of the service call, we get all tree values the service returned by using the `tpreturn` function. We will learn about each of these in *Chapter 5, Developing Servers and Clients*, of this book, but right now, we will use just the third one, containing a string converted to uppercase. Therefore, we will use an underscore for the first two values, which is a standard Python way of indicating values that have been ignored.

Creating a configuration file for Tuxedo

The easy part is over now and Python will not help us with the next step. We have to create a Tuxedo configuration file, and here is the shortest one that gets the job done. There are no restrictions in terms of the filename, but the Tuxedo documentation calls it `ubbconfig`, so we will use the same name:

```
*RESOURCES
MASTER tuxapp
MODEL SHM
IPCKEY 32769
*MACHINES
"63986d7032b1" LMID=tuxapp
    TUXCONFIG="/home/oracle/code/02/tuxconfig"
```

```
    TUXDIR="/home/oracle/tuxhome/tuxedo12.2.2.0.0"
    APPDIR="/home/oracle/code/02"
*GROUPS
GROUP1 LMID=tuxapp GRPNO=1
*SERVERS
"toupper.py" SRVGRP=GROUP1 SRVID=1
    REPLYQ=Y MAXGEN=2 RESTART=Y GRACE=0
    CLOPT="-s TOUPPER:PY"
```

A Tuxedo configuration file consists of several sections. Each section starts with an asterisk ("*") and is followed by the section name. In the following sections, we will cover each configuration section in detail.

RESOURCES section

The RESOURCES section contains parameter names and values for the whole application. Each parameter is specified on a separate line. Here are the most important values no Tuxedo application can exist without:

- MASTER: This gives the names of machines from the MACHINES section that has the master copy of the Tuxedo configuration file (TUXCONFIG). All examples in this book and many Tuxedo applications have only a single machine.

- MODEL: This typically has SHM as its value, and it stands for **shared memory**. The other option is MP, for multimachine configuration, but this is rarely used.

- IPCKEY: This is a numeric value greater than 33,768 and less than 262,143. It is something specific to Tuxedo's use of System V IPC resources and gives the address of shared memory that all Tuxedo processes connect to. Since we will be running examples in a Docker container, a single constant value is fine. On a real server running multiple applications, a unique IPCKEY key must be assigned to each application.

There are more resources you can configure here, and you can read about them in the Tuxedo documentation (https://docs.oracle.com/cd/E72452_01/tuxedo/docs1222/rf5/rf5.html#3370051), but this is sufficient for our application.

MACHINES section

This section contains configurations for all logical machines in the Tuxedo application. It is common to have just one machine and achieve a distributed application through other mechanisms, such as Tuxedo domains.

Each machine entry begins with a physical machine name, which, on Unix, matches the output of the `uname  -n` command and matches the hostname of the server. This name is limited to 30 characters in Tuxedo, and sometimes the hostname has to be shortened for continuous integration systems and others.

The machine name is followed by parameters, but this time the name and value are separated by an equals sign and multiple parameters can be given on the same line. The most important parameters are the following:

- `LMID`: This is the unique machine identifier that will be used instead of the machine name in the rest of the configuration file. We will use `tuxapp` for that.

- `TUXDIR`: This is the absolute pathname of the directory where we have installed Tuxedo. For our Docker image, it is `/home/oracle/tuxhome/tuxedo12.2.2.0.0`.

- `APPDIR`: This is the absolute pathname of the directory where you have your Tuxedo servers, or `toupper.py` in this case. For me, it is the `/home/oracle/code/02` directory. Multiple directories can be specified separated by a colon, just like Unix path variables if you want to structure a larger application.

- `TUXCONFIG`: This is the absolute path to the binary configuration file. For a real application, you should come up with a good location for this file. But for this example, I have put the configuration file along with the Tuxedo server and the `TUXCONFIG` value in `/home/oracle/code/02/tuxconfig`. The value of this parameter must be the same as the `TUXCONFIG` environment variable in order for many tools and our application to work.

GROUPS section

All Tuxedo servers are collected into groups and this section contains the group's configuration and the default values inherited by servers belonging to this group. While it does not serve any purpose for our application, a group configuration is still required. The format of a group configuration is similar to that of a machine configuration.

Each entry begins with a group name that must be unique and 30 characters or fewer in length. We will use `GROUP1` as the name.

The group name is followed by a list of parameters, and we will provide just the bare minimum:

- GRPNO: This is a unique group number larger than 0 and less than 30,000. Yes, having a unique group name is not enough and we must assign a unique number as well. This number is used instead of a group name internally, as we will see in the *Administering the application* section.

- LMID is the logical machine identifier the group resides on. For our application, it is tuxapp.

SERVERS section

Finally, we have reached the section where we describe our Tuxedo servers.

Each entry begins with the name of the executable file, which is toupper.py in our case. Tuxedo will look for a file with this name in the directories given in the APPDIR configuration parameter and then continue looking in all directories given in the Unix PATH environment variable until it finds it.

The server name is followed by a list of parameters, just like for machines and groups:

- SRVGRP: This is the name of the group a server belongs to; in our case, it is GROUP1.

- SRVID: This is a unique server identifier larger than 0 and less than 30,000. Technically speaking, this value must be unique within the group, and servers in different groups can have the same SRVID value. However, I suggest keeping SRVID globally unique, regardless of the group, as it simplifies administrative commands.

- REPLYQ: With the value Y, this will create a separate queue for service replies instead of reusing the request queue. The default value for this parameter is N, but there is no good reason for that, except maybe saving a few kilobytes of memory on servers manufactured in the 1980s.

- RESTART: This, again, is one of the parameters that must be set to "Y" instead of the default "N" because we want the executable to be restarted in case of a crash.

- MAXGEN: This should be set to a value *greater than* 1 and less than 256. The default value of 1 means that the executable will be started once but not restarted, which is not what you would expect as good default behavior. I think 2 is a good value in combination with the next GRACE parameter.

- GRACE: This is the number of seconds during which the server can be restarted MAXGEN times minus one. A valid value is equal to or greater than 0 and less than 2,147,483,648, with a default value of the number of seconds in 24 hours. Unless you have good reason for not doing so, set GRACE to 0 and you will get infinite restarts. The only downside is that if the process crashes and produces a core dump, it may quickly fill up the whole available disk space.

- CLOPT: This has to be set to -s TOUPPER:PY for our application, but it deserves a more detailed description.

CLOPT stands for *command-line options*, and it consists of two parts separated by two dashes (--), the first part recognized and processed by the Tuxedo runtime system, and the last part ignored by the Tuxedo runtime system but processed by application code. Application code receives everything specified in CLOPT and is responsible for distinguishing both parts. Here are the most common options recognized by Tuxedo:

- -s SERVICE:ROUTINE offers a service named SERVICE that will be processed by a routine named ROUTINE. The tuxedo module we used to develop the server has a single routine named PY, and we expose it as a service, TOUPPER. That gives us the value of CLOPT -s TOUPPER:PY.

- -A offers all the services given at compilation time and is typical for servers written in C.

- -s SERVICE1,SERVICE2 is also typical for servers written in C and says to offer only specific services, SERVICE1 and SERVICE2.

That is enough for this chapter, and we will continue learning about new parameters as they are needed in future chapters.

TUXCONFIG

TUXCONFIG has a special role in Tuxedo. The environment variable with this name points to the file containing the binary configuration of the Tuxedo application. We must export the TUXCONFIG environment variable with the same value as in the ubbconfig file's MACHINES section. All Tuxedo tools and application servers use this environment variable to discover the configuration file and get all the information needed from there. If you followed all the preceding steps by the letter, this command will do that for you:

```
export TUXCONFIG=`pwd`/tuxconfig
```

Now we can use the `tmloadcf` tool to compile the textual representation of configurations into the binary format used by Tuxedo applications. I prefer to use `tmloadcf` with the `-y` option, which avoids the interactive question of "Do you want to do this?", and the last option is a filename of the textual configuration file:

```
tmloadcf -y ubbconfig
```

You will get a couple of warnings about some important parameters missing, but that is fine for our application, and default values are sufficient. You should see a newly created `tuxconfig` file. If you look at the file sizes of `tuxconfig` and `ubbconfig`, the binary representation is much larger. The binary file contains space reserved for the maximum number of servers, services, and other file entries.

> **Tip**
>
> An important thing to keep in mind is that `tmloadcf` works only when the application is not running. It will give an error if you try. If you're interested in changing application configurations at runtime, we will look at this in *Chapter 6, Administering the Application Using MIBs*, of this book.

There is also a tool that does the reverse of `tmloadcf` and prints the binary configuration file in a textual representation. It is useful for getting the configuration of a running application. It also returns all supported configuration parameters filled with the default values. Just type the following command:

```
tmunloadcf
```

But we have already spent too much time on configuration and it's now time to finally run the application.

Starting the application

Following all this preparation, we can boot the application by using the `tmboot` tool with my favorite `-y` option:

```
tmboot -y
```

You will see the following result:

```
Booting all admin and server processes in /home/oracle/code/02/
tuxconfig
INFO: Oracle Tuxedo, Version 12.2.2.0.0, 64-bit, Patch Level
(none)
```

```
Booting admin processes ...
exec BBL -A :
        process id=1296 ... Started.
Booting server processes ...
exec toupper.py -s TOUPPER:PY :
        process id=1297 ... Started.
2 processes started.
```

That's it! Our application is finally up and running and we can run our client program against it. Just run the following command:

```
python3 client.py
```

A HELLO, WORLD! message in uppercase will be printed to the console.

Administering the application

Tuxedo comes with the tmadmin tool for administering the application. But there are a couple of things that can be done without it. First, let's look at the processes started by running tmboot. A common trick for distinguishing Tuxedo processes from non-Tuxedo processes is to look for ULOG in the process options (unless you have specified a different ULOGPFX configuration parameter):

```
ps aux | grep ULOG
```

This command will print the two processes started, BBL and python3, which run our processes. Notice how many command-line options the running Tuxedo process is given compared to what was printed to the console. If you are curious, use the following command:

```
tmboot -n -d1
```

This will print the real commands that Tuxedo's tmboot uses to start the application. It turns out that tmboot is not special and performs no magic: the entire application can be started by hand using the real commands from the output. tmboot just runs executables with those command-line options.

A special command-line option is -v, which prints information about services that the executable contains and routines that implement those services. Try it on our server:

```
./toupper.py -v
```

It will print : PY, meaning that servers in Python do not have any services specified at compile time, and that it contains a single routine, PY, for handling service requests. The output is more interesting for servers written in C; just try it on your existing application!

You will see a file with the ULOG prefix and the current date as the suffix, like ULOG.121220 in my case. That is the main log file of the Tuxedo application, and normally should only contain the lines about starting and stopping servers. But you must keep an eye on this file in case application errors occur. To get more information about ULOG, you can turn on Tuxedo trace by using the following command:

```
echo chtr "*:ulog:dye" | tmadmin
```

Try and run our client program (python3 client.py) and see what additional information is written in the log file. That is a good way to understand what the application does, but it should not be used on production servers. To turn trace off, use the following command:

```
echo chtr off | tmadmin
```

Sometimes, it is important to ascertain the version and patch level information of the Tuxedo runtime. The patch level is important as Oracle increases it for each bugfix instead of the version number and you might need to provide it for service requests. Here is a command to print the version information:

```
tmadmin -v
```

Sometimes it is useful to look at the overall application information, called the Bulletin Board status:

```
echo bbs | tmadmin
```

The next useful command is to print information regarding all the servers:

```
echo psr | tmadmin
```

It will print information about the two processes we saw earlier, as well as some statistics about service requests processed by the server in the RqDone column. You can run our client program and see how the number in the RqDone column changes. Another interesting fact to notice is that the BBL admin process is just another Tuxedo server.

To print service information, use the following command:

```
echo psc | tmadmin
```

This will show that the TOUPPER service is implemented by the PY routine, just as we specified in the configuration file. We can also see the number of service requests processed in the # Done column.

For application processing the messages all the time, it is also highly recommended to check the queue status by using the following command:

```
echo pq | tmadmin
```

In our case, the message queues will be empty. However, if messages are queuing up for your production application, you must check for performance problems or reconfigure the application to handle more loads.

Stopping the application

Now we can shut down the application by using the following command:

```
tmshutdown -y
```

This will stop all Tuxedo servers in the reverse order to how they were started, first our Python server and then the BBL admin process. Sometimes, however, a graceful shutdown does not succeed and tmshutdown will ask whether you want to kill the hanging process. It is fine to kill servers on a test application, but this requires more thought on production servers. There is also a final resort to shut down the application by brutally removing all queues and shared memory used by the application. Use it wisely:

```
tmipcrm -y
```

At this point, we have completed the full application life cycle. No part of the application is running after this and all memory, semaphores, and queues are released back to the operating system.

> **Tip**
>
> For the duration of this book, get into the habit of loading the configuration (tmloadcf -y ubbconfig), starting the application (tmboot -y), doing exercises, and stopping the application (tmshutdown -y). And then doing the same steps over again. There will be a lot of examples and we will save bytes by not mentioning these commands.

Summary

In this chapter, we created a simple Tuxedo server and client. We also created the Tuxedo configuration file for our application. It is a tedious task even if we do only the necessary and most important parts. After that, we started and tested our first application, and finally, we inspected the running application using both Unix standard tools and Tuxedo administration tools.

Real-world Tuxedo applications are more complex than this, but now you know how to start and stop them and where to look in terms of what services, servers, and configurations make them tick.

Before we start writing more code, we will learn some implementation and configuration aspects of Tuxedo in the next chapter using the application we created in this chapter.

Questions

1.   What environment variable contains the full path to Tuxedo's configuration file?
2.   What command is used to compile the textual configuration into the binary format used by Tuxedo?
3.   What command is used to start a Tuxedo application?
4.   What command is used to stop a Tuxedo application?
5.   What command can you use to list the processes of a Tuxedo application?

Further reading

- *Getting Started with Oracle Tuxedo* covers configuration files and administrative tools in more detail.

3
Tuxedo in Detail

There are a couple of Tuxedo's implementation details that it is useful to know when it comes to making the correct technical decisions. One of them is the BBL process present in all Tuxedo applications, and the other is the use of Unix System V IPC queues. Monitoring and sizing Tuxedo's queues is one of the most important tasks for the administrator of a Tuxedo application.

In this chapter, we will cover the following topics:

- Learning what BBL is
- Understanding queues
- Introducing services and load balancing

By the end of this chapter, you will have practical experience of how BBL contributes to application availability. You will understand the relationship between queues and services and the ways in which that can be expressed in the Tuxedo configuration and how load balancing works. You will also know how to get information about queue size and how to change it.

Technical requirements

All the code for this chapter is available at `https://github.com/PacktPublishing/Modernizing-Oracle-Tuxedo-Applications-with-Python/tree/main/Chapter03`.

The Code in Action video for the chapter can be found at `https://bit.ly/2NrZptR`.

Learning what BBL is

Before diving into a more specific topic, we should demystify the BBL process. Tuxedo uses all three System V interprocess communication mechanisms: message queues as a transport mechanism, semaphores for synchronization, and shared memory for keeping application configuration and dynamic information about the state of the application.

The shared memory region is called the **Bulletin Board** and the administration process that runs in each Tuxedo application is called the **Bulletin Board Liaison** (**BBL**). Because the Tuxedo application consists of many Unix processes starting, working, stopping, and sometimes crashing at different moments in time, it is important to keep the Bulletin Board up to date and ensure consistency despite misbehaving clients and servers.

Monitoring server processes and restarting them as needed is one of several tasks assigned to the BBL process. To investigate, we start with the Tuxedo application developed in the previous chapter and make a couple of adjustments to the configuration file I have highlighted:

```
*RESOURCES
MASTER tuxapp
MODEL SHM
IPCKEY 32769
SCANUNIT          10
SANITYSCAN         1
*MACHINES
"63986d7032b1" LMID=tuxapp
    TUXCONFIG="/home/oracle/code/03/tuxconfig"
    TUXDIR="/home/oracle/tuxhome/tuxedo12.2.2.0.0"
    APPDIR="/home/oracle/code/03"
*GROUPS
GROUP1 LMID=tuxapp GRPNO=1
*SERVERS
"toupper.py" SRVGRP=GROUP1 SRVID=1
    REPLYQ=Y MAXGEN=2 RESTART=Y GRACE=0
    CLOPT="-s TOUPPER:PY"
```

Additional parameters in the RESOURCES section have the following meaning:

- SCANUNIT: This is the number of seconds between periodic scans executed by the BBL process, and it must be a multiple of 2 or 5. The latest versions of Tuxedo allow specification in milliseconds by adding an "MS" suffix and a number between 1 MS and 30,000 MS. We will use the default value of 10 for our examples to give us plenty of time to observe different behavior.

- SANITYSCAN: This is a multiplier for the interval between periodic sanity checks of Bulletin Board data structures and server processes. We use the value of 1, which provides a sanity check every 10 seconds (SCANUNIT * SANITYSCAN).

We can now load the configuration and start the application, as we learned in the previous chapter. To demonstrate sanity checks, we will perform the following steps:

1. List all processes of the Tuxedo application. This should include BBL and python3.

2. Terminate our Tuxedo server.

3. Ensure that the server process is absent.

4. Wait for 10 seconds for BBL to perform a sanity check and restart the server.

5. Ensure that the python3 server process has been restarted by BBL.

The commands for doing this are as follows:

```
ps aux | grep ULOG
kill -s SIGSEGV `pidof python3`
ps aux | grep ULOG
sleep 10
ps aux | grep ULOG
```

It is important to remember that a process terminated with the SIGTERM signal will not be restarted by Tuxedo because it is considered a graceful shutdown. The kill command uses SIGTERM by default and we use it with the SIGSEGV signal, which is typical for memory-related errors.

There are other duties that BBL performs, and we will discover these gradually in the chapters that follow, but first, we must understand the queue mechanisms used by Tuxedo.

Understanding queues

As we learned in *Chapter 1, Introduction and Installing Tuxedo*, parts of Tuxedo's APIs were standardized in XATMI specifications with the hope that developing applications according to XATMI specifications would lead to vendor-neutral solutions and it would be easy to port applications from one XATMI implementation to another. This idea looks good on paper, but no specification can cover all aspects of behavior and be abstract at the same time. No surprises there.

The server provides resources that the client can access. A server can also act as a client and ask for the resources provided by another server as shown in the following diagram. The XATMI specification does not say how the request and response should be delivered between the client and server:

Figure 3.1 – Client-server model

Tuxedo uses queues for that. Servers have a request queue where they receive requests from clients. Servers and clients have a response queue where responses from servers are processed as shown in the following diagram. There is a configuration parameter (REPLYQ=N) that allows the server to use a single queue for both requests and replies, but I suggest avoiding this feature:

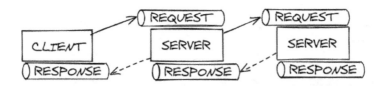

Figure 3.2 – Client-server model with queues

The configuration file of our first application did not mention any queues, so where do they come from?

Configuring queues

Tuxedo is implemented on top of Unix System V IPC queues and that explains some of the behavior and features Tuxedo has. For starters, a System V IPC queue API does not support timeout for enqueue and dequeue operations (unlike the newer POSIX queue API) and Tuxedo has to implement operation timeouts using other mechanisms. We will learn about this in *Chapter 7, Distributed Transactions*.

System V IPC queues come with three configuration parameters. To find out the current values, use the following command:

```
sudo sysctl -a | grep kernel.msgm
```

On the Docker image we created, it will display the following values:

```
kernel.msgmax = 8192
kernel.msgmnb = 16384
kernel.msgmni = 32000
```

The meaning of each cryptic parameter is as follows:

- `msgmax`: This is the maximum size of an individual message.

- `msgmnb`: This is the maximum number of bytes in a single queue.

- `msgmni`: This is the maximum number of queues in the system.

However, Tuxedo can work around these limits and transfer tens, if not hundreds, of megabytes of data even if the message size exceeds the queue size or the individual message size. This technique is called **file transfer**. Under certain conditions, instead of storing the whole message in the queue, Tuxedo will store the message in a temporary file and put a tiny message in the queue containing the filename as shown:

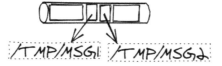

Figure 3.3 – Message transport

It is hard to tell how many messages fit in the queue because Tuxedo may choose between storing the whole message in the queue or the file under two conditions:

- The message size is larger than the queue size or message size limit.

- The message size is lower than the message size limit, but there is insufficient space in the queue due to messages the queue already contains.

However, file transfer comes at a price: storing a message on a physical filesystem is slower than storing it in the queue. To improve performance, you must make sure that most of the messages in your application do not exceed the message size maximum and that the queue size is big enough to contain all messages. A properly tuned system should never fill its queues. We will develop tools to monitor queues in *Chapter 6, Administering the Application Using MIBs*.

For example, to support messages with a size of up to 32 Kb and queues to accommodate 100 such messages, you should do the following:

```
sudo sysctl -w kernel.msgmax=32768
sudo sysctl -w kernel.msgmnb=3276800
```

Unfortunately, the whole Tuxedo application has to be restarted in order for changes to take effect and therefore it is better done ahead of time when preparing the environment for the application.

The main disadvantage of message-oriented middleware is that it requires an extra component called a **message broker**, which can introduce performance and reliability issues. Tuxedo is **broker-less** in this aspect because queues live in the OS kernel, as the following diagram demonstrates:

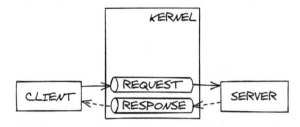

Figure 3.4 – Kernel queues

But if Tuxedo does not have a message broker, how are other message broker's tasks performed, such as routing the message to one or more destinations, load balancing, and location transparency?

Introducing services and load balancing

Unlike other messaging-oriented middlewares that operate with queues, Tuxedo operates with another abstraction level on top of that – a service. A service may be backed by one or more queues and multiple services may use the same queue. The mapping between service names and queues is stored in the Bulletin Board. The client uses this mapping information and chooses the appropriate queue and tells the kernel which queue to use for a request, as shown in the following diagram. No mapping is required for the response queue as the client tells the server exactly which queue to use for the response:

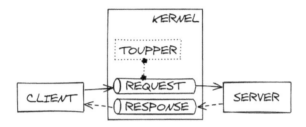

Figure 3.5 – A service

There are a couple of ways in which the service abstraction over queues can be used, so let's explore them in detail.

Exploring the MSSQ configuration

MSSQ stands for **Multiple Servers Single Queue**. In modern terminology, we would say that a queue has multiple consumers. Instead of each server having its request queue, multiple servers can share the same request queue. The main advantage is that any idle server can pick the next request from the queue and improve latency. The opposite of that would be a full queue for a single server struggling to complete all the requests, while other servers are idle with their request queues empty. Another advantage is that a server can crash or can be upgraded to a new version without any interruptions or downtime for the client because other servers continue processing requests.

Multiple servers wait for the request to arrive on the same request queue, and the first one that receives the message processes it and returns a response, as shown in the following diagram:

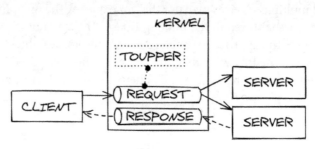

Figure 3.6 – Multiple Servers Single Queue

To see how it works, let's use the application we developed in the previous chapter and modify the configuration file:

```
*RESOURCES
MASTER tuxapp
MODEL SHM
IPCKEY 32769
*MACHINES
"63986d7032b1" LMID=tuxapp
    TUXCONFIG="/home/oracle/code/03/tuxconfig"
    TUXDIR="/home/oracle/tuxhome/tuxedo12.2.2.0.0"
    APPDIR="/home/oracle/code/03"
*GROUPS
GROUP1 LMID=tuxapp GRPNO=1
*SERVERS
"toupper.py" SRVGRP=GROUP1 SRVID=1
    REPLYQ=Y MAXGEN=2 RESTART=Y GRACE=0
    CLOPT="-s TOUPPER:PY"
    MIN=2 MAX=2
    RQADDR="toupper"
```

The main change is that we start two copies of the server instead of one. One way to achieve this would be to copy the configuration of the first server and duplicate it with a new unique SRVID. But there is a better way to do this in Tuxedo:

- MIN: This is the number of copies of the server to start automatically. The default value is 1. If this number is greater than 1, the server identifier, SRVID, will be assigned automatically up to SRVID + MAX - 1.

- MAX: This is the maximum number of server copies that can be started. It defaults to the same value as MIN. All copies above MIN must be started manually.

- RQADDR: This is the symbolic queue name and allows multiple servers to use the same physical queue. If it is not given, GRPNO.SRVID is used to create a unique symbolic queue name.

After that, we will change the client.py program to call the service 100 times:

```
import tuxedo as t
for _ in range(100):
    _, _, res = t.tpcall("TOUPPER", "Hello, world!")
    print(res)
```

Then we load the new configuration and start the application to see that it starts two instances of our Tuxedo server and that both instances use the same queue name:

```
echo psr | tmadmin
```

After that, we can run the client program and observe how the requests completed (RqDone) column has changed:

```
python3 client.py
echo psr | tmadmin
```

You should see that each server instance has processed 50 requests or a number close to that. And that is exactly what we expected: requests are evenly distributed among server instances. If you feel adventurous, you can modify the client with an infinite loop, run it, kill one of the server instances, and see whether you still get the responses.

In the previous output, you will also notice a mysterious value around 2,500 as Load Done, but that is better explained with the next configuration.

Exploring the SSSQ configuration

SSSQ stands for **Single Server Single Queue**. Each server has its request queue in this setup. Since Tuxedo is broker-less, it is the responsibility of the client to perform load balancing and choose the appropriate queues for the request. While it may seem like a step back from the MSSQ configuration, SSSQ has its pros as well:

- It is recommended when messages are big and quickly fill the queue. Having multiple queues in such a scenario is better.

- SSSQ is better when you have a large number of servers. MSSQ may experience contention and perform worse with more than 10 servers on a single queue.

- SSSQ can offer unique services in addition to the common ones. All servers in the MSSQ configuration must provide the same services.

Each server waits for the request on its request queue and the client chooses the most appropriate request queue according to the load balancing algorithm. Refer to the following diagram:

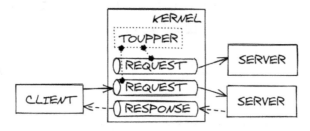

Figure 3.7 – Single Server Single Queue

The configuration file for an SSSQ setup is as follows:

```
*RESOURCES
MASTER tuxapp
MODEL SHM
IPCKEY 32769
LDBAL Y
*MACHINES
"63986d7032b1" LMID=tuxapp
     TUXCONFIG="/home/oracle/code/03/tuxconfig"
     TUXDIR="/home/oracle/tuxhome/tuxedo12.2.2.0.0"
     APPDIR="/home/oracle/code/03"
*GROUPS
```

```
GROUP1 LMID=tuxapp GRPNO=1
*SERVERS
"toupper.py" SRVGRP=GROUP1 SRVID=1
    REPLYQ=Y MAXGEN=2 RESTART=Y GRACE=0
    CLOPT="-s TOUPPER:PY"
    MIN=2 MAX=2
```

In addition to using MIN and MAX, which we learned about previously, we skip RQADDR and let Tuxedo assign a unique queue for each server. The most important addition is as follows:

- LDBAL: This turns load balancing between the queues on (Y) or off (N). It is off by default because of some perceived performance overhead due to additional calculations needed, but I have not observed it in practice. More importantly, without load balancing, Tuxedo will pick the first queue and fail to use any others.

Tuxedo's load balancing algorithm is described in detail in a white paper published by Oracle: https://www.oracle.com/technetwork/middleware/tuxedo/overview/ld-balc-in-oracle-tux-atmi-apps-1721269.pdf. In short, it assigns each service a weight, calculates how much each queue weighs, and picks the lightest one. Unfortunately, it does not take into account the size of the message and the space available in the queue, so it may choose the queue where a much slower file transfer is needed.

We can load the new configuration and start the application to see that it starts two instances of our Tuxedo server and that each instance uses a different queue name:

```
echo psr | tmadmin
```

After that, we can run the client program that executes 100 requests and observe how the requests completed (RqDone) column has changed:

```
python3 client.py
echo psr | tmadmin
```

You should see that only one server instance has processed all 100 requests or a number close to that. Why did the second instance of the server not process any requests? The weight is assigned to the queue when the request is put into it and decreased once the request is retrieved from the queue. Since we serially call services one after another instead of in parallel, the queue's weight is 0 when load balancing is performed and the first queue is picked up. We can launch several clients in parallel to utilize the second instance of the server:

```
python3 client.py & python3 client.py & python3 client.py
echo psr | tmadmin
```

But even now, you will see that the load is not evenly distributed among server instances.

As there are pros and cons to both configurations, it is not easy to choose one for your application. My experience with a specific application shows that a hybrid approach works best: multiple queues, each in an MSSQ setup, combining the best properties of both. Oracle's TPC-C benchmark setup uses SSSQ mode. Finding the best setup for your application and hardware can only be ascertained by multiple iterations of tuning the configuration and measuring the result.

Summary

In this chapter, we configured the Tuxedo application's sanity checks and witnessed how it restarts crashed processes. We learned about sizing queues and how those queues are used to deliver requests and responses between the client and server. Finally, we learned about services and their relationship with queues by modifying the configuration and examining the application behavior.

We have learned a lot about the inner workings of Tuxedo and message queues, but the message itself and the data contained in queues remain abstract and mysterious. There are specific message data types that Tuxedo supports and we will learn about these in the next chapter.

Questions

1. What termination signal will cause BBL to not restart the server?

2. Is there a direct limit in terms of messages in the queue?

3. What is the Tuxedo term for a single queue with multiple consumers?

4. What is the Tuxedo term for a single queue with a single consumer?

Further reading

- Tuxedo documentation at `https://docs.oracle.com/cd/E72452_01/tuxedo/docs1222/`.

4
Understanding Typed Buffers

The client-server model allows the client to access resources provided by the server. We already learned that communication is accomplished by sending messages through queues. The messages exchanged in Tuxedo are described in the XATMI specification, and Tuxedo has implemented more message types since. These messages are called **typed buffers** and you can't have a Tuxedo application without them. In this chapter, we will cover the following topics:

- Introducing typed buffers
- Using STRING typed buffers
- Using CARRAY typed buffers
- Using FML32 typed buffers
- Importing and exporting typed buffers

By the end of this chapter, you will have a good understanding of the typed buffers available from Python code. We will explore the features and most common pitfalls of FML32. You will learn how to persist messages and build some useful tools along the way.

Technical requirements

The code for the chapter can be found here:

`https://github.com/PacktPublishing/Modernizing-Oracle-Tuxedo-Applications-with-Python/tree/main/Chapter04`

The Code in Action video for the chapter can be found at `https://bit.ly/38SSscG`.

Introducing typed buffers

Today the most common message exchange formats are JSON and XML, and the most widely used protocol (HTTP) has a special header called **content-type** to describe the payload type. Tuxedo was created at a time when XML and JSON were not invented yet. At this point, for a C program to carry data, along with metadata describing its format, was an innovative idea. And that is what a typed buffer is: data with a format (type) description.

There are XATMI standard types called `STRING` and `X_OCTET` (`CARRAY`) for zero-terminated C strings and binary data. There are non-standard `FML` and `FML32` types implemented by Tuxedo. We will learn more about them later in this chapter.

However, several typed buffers are not supported by the Python library. I do not suggest using these types in new applications, but if you do use them in existing applications you will have to either update the Python libraries or implement a conversion to the supported types. These types are as follows:

- `MBSTRING`: This is a character array of multibyte characters similar to the `STRING` type.

- `VIEW, VIEW, X_C_TYPE, X_COMMON`: These are different variations of C structures (`struct`) with specific rules for encoding arrays and `NULL` values.

- `RECORD`: This is a COBOL copybook record. You can think of it as the COBOL version of the `VIEW` type.

- `XML`: This is an XML document. Tuxedo knows that it is an XML document and can convert between ASCII and EBCDIC and do data-dependent routing. To the developer, it does not look much different from a regular string.

Next, let's look at the types supported by the Python library.

Using STRING typed buffers

STRING is a C-style array of characters terminated with a NULL character. This is the typed buffer we used in all previous examples for converting a string into uppercase. You should know by now that it is a regular Python str type and the library converts it to and from the typed buffer under the hood.

Writing clients and servers that exchange strings are a good first example but not enough for the real-world requirements. You can exchange *strings on steroids* by sending data serialized to JSON or XML but you will find a better alternative later in this chapter.

Using CARRAY typed buffers

CARRAY is an array of characters that may contain NULL characters. It is also known as an X_OCTET typed buffer in the XATMI specification. Unlike a STRING typed buffer, you can use CARRAY to exchange any binary data. Or you can serialize structured data using Protocol Buffers, Apache Thrift, Apache Avro, or similar methods to achieve a compact size, extensibility, and support different programming languages.

bytes is the Python type that is automatically converted to and from the CARRAY typed buffer. Just by changing our client program, we would send CARRAY to the service:

```
import tuxedo as t
_, _, res = t.tpcall("TOUPPER", "Hello, world!".encode("utf8"))
print(res)
```

Of course, the server will receive bytes instead of str, and malfunction, but you get the idea of how to move forward. We will not spend any time on that because it is now time to introduce the one true typed buffer you should use for all applications.

Using FML32 typed buffers

FML stands for **Field Manipulation Language**. Do not be confused by the language part, FML is a data structure containing field identifiers and field values in a flat, contiguous memory block. It is similar to a *multimap* or *multidict* in other programming languages where more than one value may be associated with a given key. It also supports many different value types in the same data structure including FML32 containing other FML32 buffers. FML is an older data structure that supports 16-bit long identifiers, number of fields, and value lengths. FML32 raises those limits and 32 bytes are used for field identifiers, number of fields, and value lengths. All modern applications should use FML32.

If my words are not enough, another reason in favor of FML32 is that it is used by Tuxedo internally:

- Management services and APIs work with the FML32 type.

- Request metadata uses FML32.

- Jolt and other features use and support FML32.

- Tuxedo comes with several tools for manipulation of FML32, unlike other types.

FML32 contains field identifiers and values. But using 32-bit identifiers in the code and trying to remember what each of them stands for is complicated. That is why Tuxedo provides utility functions for maintaining a bidirectional mapping between field identifiers and field names. Strictly speaking, having field names is not required but they make application development easier and enable other features, such as Boolean expressions, which we will explore later.

Two environment variables are needed for bidirectional mapping between field names and field identifiers:

- FIELDTBLS32: This is a comma-separated list of field table files. The file can be on an absolute path or a relative path in which case it is searched for in directories specified in FLDTBLDIR32.

- FLDTBLDIR32: This contains a list of directories separated by a colon to look for field table files. An empty FLDTBLDIR32 causes files to be searched for in the current directory.

Several field definitions come with Tuxedo and are useful for everyday tasks. These files can be found in the $TUXDIR/udataobj directory:

- tpadm: This contains MIB fields for Tuxedo administration using the .TMIB service or the tpadmcall function.

- Usysfl32: This contains a definition for the SRVCNM field that is needed for ud32 to specify the service name.

If you work with Tuxedo domains, **Service Architecture Leveraging Tuxedo (SALT)**, you might want to include other files from this directory. The environment variables will also contain filenames and directories of the field definition tables of your application, but for the examples in this book, we will add the current folder and `example` as the filename for our field definitions. We will create the `example` file later in the chapter. The commands to add new environment variables are as follows:

```
export FLDTBLDIR32=$TUXDIR/udataobj:`pwd`
export FIELDTBLS32=Usysfl32,tpadm,example
```

To give you a sense of the power of `FML32`, execute the following command:

```
echo -e "SRVCNM\t.TMIB\nTA_CLASS\tT_DOMAIN\nTA_OPERATION\
tGET\n" | ud32
```

The preceding command will print the information about your application, similar to the one you see in the `*RESOURCES` section when unloading the Tuxedo configuration using the `tmunloadcf` tool. What you accomplished with the command was a call to the `.TMIB` service provided by the `BBL` server and passed two fields: `TA_CLASS` with `T_DOMAIN` value, and `TA_OPERATION` with the `GET` value. You passed also the third field, `SRVCNM`, with the `.TMIB` value but that was used by the `ud32` tool, not the service itself.

We just called a service without writing any code at all! But this book is about programming in Python so let's do the same using Python!

Using FML32 in Python

Unlike some other wrappers of Tuxedo that expose all of the existing `FML32` manipulation API, the Python `tuxedo` module (and its inspiration, `tuxmodule`) converts the whole `FML32` message to a Python `dict` type and converts it back to `FML32` when needed. That saves us from learning and memorizing all of the different `FML32` manipulation functions and their quirks in favor of using standard Python types. Just look at the list of functions in the Tuxedo documentation. Refer to the *Further reading* section at the end of the chapter) and you will feel relieved. I must admit, converting the whole message back and forth may sound like overkill for a simple service that changes one out of the thousands of fields present in the message. But it is pretty fast anyway, and if you care about such tiny details, you should not be using Python in the first place. The largest benefit of this approach is that most of the code will be independent of Tuxedo and easily unit-tested without starting the Tuxedo application.

FML32 in Python is a dictionary with field names as keys (str type) and lists (list type) of different types as values. As a shorthand, a single value given directly is equal to a list containing a single value. The following two Python dictionaries are converted to the same FML32 buffer:

```
{"TA_CLASS": "Single value"}
{"TA_CLASS": ["Single value"]}
```

However, when converted from FML32 to the Python dictionary it will always contain lists as values:

```
{"TA_CLASS": ["Single value"]}
```

Knowing that we can write a Python program that calls the .TMIB service and passes the TA_CLASS and TA_OPERATION fields to it. The previous example using ud32 can be written in Python as follows:

```
import tuxedo as t
_, _, res = t.tpcall(
    ".TMIB",
    {
        "TA_CLASS": "T_DOMAIN",
        "TA_OPERATION": "GET",
    },
)
print(res)
```

The output printed on the screen will be a Python dictionary but the content will be similar to the one printed by the ud32 example.

Before we proceed to the next logical topic, I want to take you on a detour to investigate FML32.

Working with field identifiers

There are several functions for working with field identifiers that are rarely used in production code but are useful for experiments, debugging, and developing tools to aid programming in C. These functions are as follows:

- The Fldid32 function uses the field name to retrieve the 32-bit field identifier used by Tuxedo internally.

- Fldname32 is the reverse function of Fldid32: it uses the 32-bit identifier to retrieve the field name.

- Fldtype32 and Fldid32 are functions to break up the field identifier into field type and number parts. The field type is a numeric constant and the value of 5 returned in this example stands for a FLD_STRING named constant.

- Fmkfldid32 is the reverse function of Fldtype32 and Fldid32: it creates the 32-bit identifier from the field type and number parts.

Let's test these functions in practice on the TA_CLASS field by using the following Python code:

```python
import tuxedo as t
fieldid = t.Fldid32("TA_CLASS")
assert t.Fname32(fieldid) == "TA_CLASS"
assert t.Fldtype32(fieldid) == t.FLD_STRING
assert t.Fldno32(fieldid) == 6000 + 2
assert t.Fmkfldid32(t.FLD_STRING, 6000 + 2) == fieldid
```

It should be obvious that the field number and numeric type are combined to get the field identifier. Here is a fact to keep you puzzled. The same field number can be used to create two different field identifiers with different types. And since identifiers are different, both fields can be present in an FML32 buffer at the same time:

```python
assert t.Fmkfldid32(t.FLD_STRING, 1) != t.Fmkfldid32(t.FLD_
LONG, 1)
```

We have now created FML32 field identifiers programmatically but it is time to do it in a correct and maintainable way.

Defining fields

Field definitions or field table files are regular text files with a few rules:

- Blank lines and lines beginning with # are ignored.

- Lines beginning with $ are used for C code generation.

- Lines beginning with *base define the base value for field numbers.

- All other lines contain field definitions.

According to Tuxedo documentation, field definitions should have the following format of five fields separated by whitespace (space or tab characters):

```
name relative-number type flag comment
```

The meaning of these fields is the following:

- name: This is a unique field name. It should be a unique name but Tuxedo does not stop you from having a duplicate. The name should not be longer than 30 characters but Tuxedo does not enforce that and will silently truncate it to 30 characters.

- relative-number: This is added to the current value of the base to obtain the field number. The resulting field number must be within the range of 10001-30000000. But as you might guess by now, Tuxedo does not prevent you from using a field number within the reserved range of 1-10000.

- type: This is the type of the field and one of short, long, float, double, char, string, carray, mbstring, ptr, fml32, or view32.

- flag: This is reserved for future use; it is nice to put a dash here.

- comment: This is an optional field not used by any tool but you can use it instead of putting a comment on a line of its own.

That was the official file format, but in practice, just the first three fields are important and everything else is ignored. Even Oracle's developers don't bother to put a dash in the **flag** field so we can write field definitions in the following format:

```
name relative-number type whatever
```

As a practical example of this format, let's look at the TA_CLASS field we examined programmatically before. It can be found in $TUXDIR/udataobj/tpadm and looks like this:

```
#NAME           ID    TYPE    COMMENT
#----           --    ----    -------
*base 6000
TA_ATTRIBUTE    0     long    IO Field identifier of class
attribute
TA_BADFLD       1     long    O  Field identifier of field in
error
TA_CLASS        2     string  IO class name, see MIB(5) and TM_
MIB(5)
```

The field number for TA_CLASS is 6002 (2 added to the value of *base) and the type is string or 5 as the numeric constant. That should match what we got programmatically.

FML32 supports many types that resemble types in the C programming language. Instead of listing and describing them all, I prefer to divide them into two groups. These are the ones I suggest using in new applications:

- string for C-style NULL-terminated strings.
- long for integers, unless the integer is too big for 64 bits, in which case, just store it in a string.
- fml32 for nested FML32 buffers to build nested data structures.
- double for floating-point numbers.

And there are other types I suggest avoiding:

- short is a shorter integer type, but prefer using long instead.
- float is a lower-precision floating-point number, but prefer using double instead.
- char is a type for a single character, but prefer using a string of length 1 instead.
- mbstring is a string of multibyte characters, but prefer having all your strings in UTF-8 and using the string type for it.
- carray is similar to string, but it may contain binary data and binary zero, which is a terminator for string. carray is the best type for transferring binary data around but I find carray fields to be harder to use, log, and debug, so I prefer using the string type and storing data in hex representation.

- `double` is tricky. Sooner or later, somebody will store financial amounts in these fields and rounding errors will catch them out.

- `ptr`, `view32`: I have not seen a good use for this and Tuxedo does not use it according to the available field tables.

Of course, Python does not have all the types available in the C code. The supported types and their mappings to Python are as follows:

- `carray` as the Python `bytes` type

- `char`, `string` as the Python `str` type

- `short`, `long` as the Python `int` type

- `float`, `double` as the Python `float` type

- `fml32` as the Python `dict` type

Let's create a field definition in a new file called `example` with the following field definitions:

```
*base 10000
PARAM            1   fml32
NAME             2   string
VALUE            3   string
AMOUNT           4   long
CURRENCY         5   string
MERCHANT_NAME    6   string
CARD_NETWORKS    7   string
SAMENAME         10  string
SAMENAME         11  string
SAMEID           11  string
```

If you correctly exported the `FIELDTBLS32` and `FLDTBLDIR32` environment variables, the newly defined fields will work. We will test that by calling the `.TMIB` service and ensuring all fields survive the round-trip. This is not the best way to do this, but I have not taught you a better one yet:

```
import tuxedo as t
_, _, res = t.tpcall(
    ".TMIB",
    {
```

```
        "PARAM": [
            {
                "NAME": "user",
                "VALUE": "oracle",
            },
            {
                "NAME": "password",
            },
        ]
    },
)
print(res)
```

Each key in the dictionary must have the field with the same name defined. Now contrast that with a simple example with an undefined field that will fail:

```
t.tpcall(".TMIB", {"THERE_IS_NO_SUCH_FIELD": "?"})
```

Failing with an exception is a good thing. It is much worse when field definitions contain duplicate field names, such as the SAMENAME field with different field numbers, or SAMEID where different fields have the same field number. Add servers implemented in multiple programming languages to the mix and you will have a huge mess and errors that are hard to debug. Try this example:

```
assert t.Fldid32("SAMENAME") == t.Fldid32("SAMEID")
assert t.Fname32(t.Fldid32("SAMEID")) == "SAMENAME"
_, _, res = t.tpcall(".TMIB", {"SAMEID": "ID"})
print(res)
```

The ID value you passed in the SAMEID field will come back in the SAMENAME field.

Tip

Tuxedo is very relaxed about long field names, duplicate names, and field numbers. If you are developing a large complex application or building many microservices, it is a good practice to build tools for keeping a track of fields: a field name index, duplicate checks, reserving intervals for each module in advance, and so on. It is also a good idea to use a prefix for each application to avoid field conflicts just like Tuxedo does with the TA_ prefix.

Displaying FML32 typed buffers

You might wonder what the following command we used before in the *Using FML32 typed buffers* section did:

```
echo -e "SRVCNM\t.TMIB\nTA_CLASS\tT_DOMAIN\nTA_OPERATION\
tGET\n"
```

This command produces a specific textual representation of FML32 that can be both printed and read by Tuxedo API functions. It is also the format that is consumed and produced by the ud32 tool. Since the format relies on whitespace characters, I use <tab> to represent tab characters and <blank line> for empty lines:

```
FIELD_NAME<tab>field_value
FIELD_FML32<tab>
<tab>FIELD_NAME1<tab>value1
<blank line>
FIELD_FML32<tab>
<tab>FIELD_NAME1<tab>value2
<blank line>
<blank line>
```

The two functions working on this representation are as follows:

- Fextread32: This reads the textual representation from a file and returns a Python dictionary.

- Ffprint32: This prints the Python dictionary in a textual representation to a file.

By putting those two functions together we can write a simple clone of the ud32 tool that reads input, calls the service with the name provided in the SRVCNM input field, and prints the service output:

```
import sys
import tuxedo as t
req = t.Fextread32(sys.stdin)
_, _, res = t.tpcall(req["SRVCNM"][0], req)
t.Ffprint32(res, sys.stdout)
```

I named the file `ud32.py` for convenience and we can test that it works like `ud32` as follows:

```
echo -e "SRVCNM\t.TMIB\nTA_CLASS\tT_DOMAIN\nTA_OPERATION\
tGET\n" | python3 ud32.py
```

This representation is specific to `FML32`, however, there is a somewhat similar feature for all typed buffers that we will learn later in this chapter.

Writing FML32 expressions

An expression is a syntactic entity that may be evaluated to determine its value. For `FML32` expressions, the value can be either a Boolean or numeric. Expressions have access to values of fields as *variables* and can perform arithmetic and Boolean operations on them. `FML32` expressions are nowhere as powerful but are used in many Tuxedo applications to introduce some configurable behavior. This is not so useful in Python code because Python has a very powerful `eval` function for this, along with other safer alternatives.

The largest limitation is that expressions do not work well with multiple occurrences of fields and do not work with nested `FML32` at all (such as our `PARAM` field).

Fields are accessed using the following:

- `field_name[number]` accesses the specified field occurrence.
- `field_name` is shorthand for `field_name[0]`.
- `field_name[?]` matches against any occurrence of the field.

You can multiply, divide, add, and subtract values using the same operators as in Python: `*`, `/`, `%`, `+`, and `-`. You can compare values using `<`, `>`, `<=`, `>=`, `==`, and `!=` in the same way as in Python. Instead of `and` and `or` you must use `&&` and `||`. You can also match against a regular expression using `%%` or `!%` with a regular expression string given in single quotes. Let's look at some examples:

```
import tuxedo as t
txn = {
    "AMOUNT": 12300,
    "CURRENCY": "USD",
    "MERCHANT_NAME": "PACKTPUB",
    "CARD_NETWORKS": [
        "VISA",
```

```
            "MASTERCARD",
            "AMEX",
    ],
}
```

The `boolev32` function evaluates an expression on the given `FML32` buffer and returns a Boolean. We can use it to check whether the merchant name starts with `PACKT`:

```
t.Fboolev32(txn, "MERCHANT_NAME %% '^PACKT.*'")
t.Fboolev32(txn, "MERCHANT_NAME !% '^PACKT.*'")
```

The first line will evaluate to `True` and the second to `False`. We can also create more complex expressions such as the following:

```
t.Fboolev32(txn, "CURRENCY == 'USD' && CARD_NETWORKS[?] ==
'AMEX'")
```

The preceding code checks whether the currency is `USD` and one of the card networks is `AMEX`.

The `Ffloatev32` function evaluates an expression on a given `FML32` buffer and returns a number. That can be used to implement a primitive calculator that converts and amount in USD to EUR major units, for example:

```
t.Ffloatev32(txn, "AMOUNT * 0.82 / 100.")
```

That concludes the overview of `FML32` features. Not all of them are useful in Python code because Python offers more powerful tools. But they are still good to know in order to understand existing applications. Now let's look at something that works with all kinds of typed buffers.

Importing and exporting typed buffers

Every buffer type supported by Tuxedo can be imported and exported in the same way. When you are using Python, it does not sound that impressive because it is very easy to either store the Python dictionary itself or convert it to JSON, XML, or any other type. However, C does not come with batteries included and this import and export capability is often used to persist messages in files or a database.

There are two modes for the format of exported data. The first and the default one is a binary representation as the Python `bytes` type. The second format is a Base64 representation of the same bytes and it has to be explicitly turned on by using the `TPEX_STRING` flag. While the second format is a bit longer, it is more friendly for logfiles and storing them in a database. Nothing stops you from encoding the binary representation yourself, but using `TPEX_STRING` is shorter and probably faster as well. The functions for doing imports and exports are called `tpimport` and `tpexport`.

First, let's try to export a simple string that maps to Tuxedo's `STRING` typed buffer:

```
import tuxedo as t
print(t.tpexport("Hello, world!"))
```

Somewhere among all bytes, you should see a `STRING` string for the type information and `Hello, world!` for the data exported. On both sides of that, there will be bytes reserved for the headers and padding characters.

We can also export a Python dictionary that maps to the Tuxedo `FML32` typed buffer:

```
print(t.tpexport({"TA_CLASS": "Hello, world!"}))
```

Just like in the previous example, you should see an `FML32` string for the type information and `Hello world!` for the field value. What you will not see is the `TA_CLASS` field name because the typed buffer contains the field identifiers, not names.

We can also export the `FML32` typed buffer in Base64 representation:

```
print(t.tpexport({"TA_CLASS": "Hello, world!"}, t.TPEX_STRING))
```

If you feel curious, you can decode the resulting string and get the same value as the binary export.

And finally, we can test the whole round-trip of importing the previously exported typed buffer in both binary representation and Base64 representation:

```
print(t.tpimport(t.tpexport({"TA_CLASS": "Hello, world!"})))
print(
    t.tpimport(
        t.tpexport({"TA_CLASS": "Hello, world!"}, t.TPEX_
STRING),
        t.TPEX_STRING,
    )
)
```

Exporting and importing a message forces the Python type to be converted into a Tuxedo typed buffer and back. This is a good way to validate that all fields are properly defined without calling any service, as we did before.

Summary

In this chapter, we learned about different message types that Tuxedo supports and how to use them from Python. We spent a lot of time on the best and most common message type supported by Tuxedo – FML32. We learned how to call services without writing any code, and investigated different features and quirks you should be aware of when using FML32. We even built a clone of the Tuxedo native ud32 tool using just a few lines of Python. And finally, we learned how to persist Tuxedo messages in a standard way.

We have learned enough information about messages exchanged in Tuxedo for the rest of the book, and now it is time, in the next chapter, to learn more details about sending these messages between clients and servers.

Questions

1. What typed buffers are supported in Python?

2. What is the tool for calling services using FML32?

3. Which function is used to create a typed buffer from a string?

4. Which function is used to evaluate an expression and return a numeric result?

Further reading

- The Tuxedo documentation: https://docs.oracle.com/cd/E53645_01/tuxedo/docs12cr2/rf3fml/rf3fml.html

Section 2: The Good Bits

Tuxedo offers a lot of functionality in the core system and with add-ons. Some functions are obscure; some are used rarely. Here we will focus on the common subset of features that are used by all applications for solving everyday tasks.

This section has the following chapters:

- *Chapter 5, Developing Servers and Clients*
- *Chapter 6, Administering the Application Using MIBs*
- *Chapter 7, Distributed Transactions*
- *Chapter 8, Using Tuxedo Message Queue*
- *Chapter 9, Working with Oracle Database*

5
Developing Servers and Clients

We developed our first to upper Tuxedo application early in *Chapter 2, Building Our First Tuxedo Application*, of the book and then deep dived into the architecture and implementation of Tuxedo and typed buffers. There is one final topic we have to explore in depth in order to fully understand the first Tuxedo application: developing Tuxedo clients and servers themselves.

In this chapter, we will cover the following topics:

- Writing to ULOG
- Understanding server lifetime
- Advertising services
- Receiving inputs in a service
- Returning outputs from a service
- Understanding client lifetime
- Calling a service
- Joining the application

By the end of this chapter, you will know how to develop Tuxedo servers. You will know several ways to expose services to clients and how they interact with different configuration modes of Tuxedo servers. You will know how to develop Tuxedo clients to consume services in synchronous and asynchronous manners and how to build more complex patterns on top of that. Finally, you will know how to deal with applications that require user authentication.

Technical requirements

The code for the chapter can be found here:

https://github.com/PacktPublishing/Modernizing-Oracle-Tuxedo-Applications-with-Python/tree/main/Chapter05

The Code in Action video for the chapter can be found at https://bit.ly/3eR7tPT.

Writing to ULOG

By now, you should have noticed files such as ULOG.mmddyy appearing in the directory when you start the Tuxedo application. *mmddyy* consists of the current month, date, and year. So far, this log file contains only messages created by Tuxedo itself.

Tuxedo also exposes a function to write messages to this log file. The function is called userlog and can be used from both Tuxedo clients and servers. Here is an example of how to use it:

```
import tuxedo as t
t.userlog("Hello, ULOG")
```

After calling the function, a line similar to the following should be present in the *ULOG*:

```
181928.15c365dcb562!?proc.1737.2520094528.-2: Hello, ULOG
```

Using the userlog function is convenient for code examples and quick debugging but it lacks many features we take for granted. Every logging library provides features such as logging levels, formatters, filters, and many more that the *ULOG* is lacking. For a real-world application, you should leave *ULOG* for Tuxedo system messages and use Python's logging module or alternative libraries instead of userlog.

However, we will continue using `userlog` throughout this book for brevity. Let's do it right now to learn more about Tuxedo servers!

Understanding server lifetime

Once we create our server object, we give full control to the Tuxedo runtime by calling the `run` function. Tuxedo then does its magic and invokes the appropriate function for the service called by the client. In addition to services, these are the several functions that Tuxedo attempts to call during the server's lifetime:

- `tpsvrinit`: This is called during server startup and receives all command-line arguments. The arguments part is not very interesting for Python code because we can always get it from `sys.argv`, but it is present to match the **C XATMI** function with the same name. What you do in this function depends on your application, but it must return `0` to indicate success and `-1` for error. A typical task in this function for Tuxedo servers is to advertise services.

- `tpsvrdone`: This is called during server shutdown and typically cleans up connections and other resources.

- `tpsvrthrinit`: This is similar to `tpsvrinit`, but is used for multithreaded Tuxedo servers and it is called for each thread started by Tuxedo.

- `tpsvrthrdone`: This is similar to `tpsvrdone`, but is called when each thread of the multithreaded Tuxedo server is terminated.

> **Important tip**
>
> When `tpsvrinit` or `tpsvrthrinit` returns a non-zero return value, the server will fail to start and Tuxedo will not make any attempts to restart the server again. You should return a non-zero value only for permanent errors that can be solved only with manual intervention.

Let's see how this works in practice. Create a file called lifetime.py that writes messages to *ULOG* when the lifetime methods are called:

```python
#!/usr/bin/env python3
import sys
import tuxedo as t
class Server:
    def tpsvrinit(self, argv):
        t.userlog(f"Starting server {argv}")
        return 0
    def tpsvrthrinit(self, argv):
        t.userlog(f"Starting server thread {argv}")
        return 0
    def tpsvrdone(self):
        t.userlog("Stopping server")
    def tpsvrthrdone(self):
        t.userlog("Stopping server thread")
    def TOUPPER(self, req):
        return t.tpreturn(t.TPSUCCESS, 0, req.upper())
t.run(Server(), sys.argv)
```

We will need a Tuxedo configuration file with the following content:

```
*RESOURCES
MASTER tuxapp
MODEL SHM
IPCKEY 32769
*MACHINES
"15c365dcb562" LMID=tuxapp
    TUXCONFIG="/home/oracle/code/05/tuxconfig"
    TUXDIR="/home/oracle/tuxhome/tuxedo12.2.2.0.0"
    APPDIR="/home/oracle/code/05"
*GROUPS
GROUP1 LMID=tuxapp GRPNO=1
*SERVERS
"lifetime.py" SRVGRP=GROUP1 SRVID=1
    REPLYQ=Y MAXGEN=2 RESTART=Y GRACE=0
    CLOPT="-s TOUPPER:PY"
```

After loading the configuration file, start the application, check the *ULOG*, stop the application, and check the *ULOG* again. What you will see in the *ULOG* is that the `tpsvrinit` and `tpsvrdone` functions were called. There is no mention of `tpsvrthrinit` and `tpsvrthrdone`. That is because every Tuxedo server is single-threaded by default.

To make Tuxedo start multiple threads, replace the `*SERVERS` section with the following one:

```
*SERVERS
"lifetime.py" SRVGRP=GROUP1 SRVID=1
     REPLYQ=Y MAXGEN=2 RESTART=Y GRACE=0
     MINDISPATCHTHREADS=2 MAXDISPATCHTHREADS=2
     CLOPT="-s TOUPPER:PY"
```

Notice the new `MINDISPATCHTHREADS` and `MAXDISPATCHTHREADS` parameters. They specify the initial and maximum number of threads started by Tuxedo. However, in practice, it is common to set both parameters to the same value and skip wondering when and how Tuxedo will start additional threads.

You can do the load-start-stop ritual of the Tuxedo application and *ULOG* will contain calls to `tpsvrthrinit` and `tpsvrthrdone` for each of the threads. Great! But what should we do in those `init` and `done` functions besides writing to the *ULOG*? The most common task is advertising services, so let's learn more about it.

Understanding the advertising of services

By advertising services, servers expose resources that clients can consume. It registers the service name in the bulletin board and associates it with the server and server's request queue.

There are two ways to advertise services using Python (there are more in the C programming language):

- The first way is by using the Tuxedo configuration and we already did it before by using the `CLOPT` parameter in the Tuxedo configuration file.

- The second way is to do it programmatically by using the `tpadvertise` and `tpadvertisex` functions. The main benefit of this approach is that the program code can decide which services to advertise and what service name to use for it. Let's see how to try each of them:

1. First, let's use the `tpadvertise` function. It takes a single argument with a service name to advertise. We will use the code of the `tpadvertise.py` Tuxedo server included in the following code. It will advertise a service name with a `PING_` prefix followed by the `SRVID` configuration parameter that is passed to the program as the `-i` command-line argument. Functionality such as this may be useful in implementing a heartbeat service, status check service, cache refresh service, and so on:

```python3
#!/usr/bin/env python3
import sys
import tuxedo as t
import argparse
class Server:
    def tpsvrinit(self, argv):
        t.userlog(f"Starting server {argv}")
        setattr(self, "PING_1", self.UNADVERTISED)
        setattr(self, "PING_2", self.UNADVERTISED)
        parser = argparse.ArgumentParser()
        parser.add_argument("-i", dest="srvid")
        args, _ = parser.parse_known_args(argv)
        self.srvid = args.srvid
        setattr(self, f"PING_{self.srvid}", self.PING)
        t.tpadvertise(f"PING_{self.srvid}")
        return 0
    def PING(self, req):
        req["TA_DEBUG"] = "PONG!"
        return t.tpreturn(t.TPSUCCESS, 0, req)
    def UNADVERTISED(self, req):
        req["TA_DEBUG"] = "UNADVERTISED called"
        return t.tpreturn(t.TPSUCCESS, 0, req)
t.run(Server(), sys.argv)
```

During the `tpsvrinit` function, we first define service functions `PING_1` and `PING_2` using `UNADVERTISED` as an implementation. Once we know the server's `SRVID` configuration parameter, we replace the appropriate `PING_x` function with the `PING` function as implementation.

2. Replace the `*SERVERS` section of your Tuxedo configuration with the following block:

```
*SERVERS
"tpadvertise.py" SRVGRP=GROUP1 SRVID=1
    REPLYQ=Y MAXGEN=2 RESTART=Y GRACE=0
    MIN=2 MAX=2
```

3. After loading the configuration and starting the application, let's do a couple of experiments. Execute the following command:

```
echo -e "SRVCNM\tUNADVERTISED\n" | ud32
```

You should get an error with the `TPENOENT` code saying that there is no such service. We did not advertise such a service, so this error is expected.

4. Next, let's call the `PING_1` service several times by using a nice feature of ud32 that repeats the service call for each additional empty line in input:

```
echo -e "SRVCNM\tPING_1\n\n\n\n\n" | ud32
```

You should get five responses from the service saying `PONG!`. Let's now do the same with the `PING_2` service:

```
echo -e "SRVCNM\tPING_2\n\n\n\n\n" | ud32
```

Again, you should get five responses. What is the point of calling the service five times?

So far, we have looked at the SSSQ configuration we learned about in *Chapter 3, Tuxedo in Detail*. Let's now try the MSSQ configuration and observe the difference.

To do that, replace the `*SERVERS` section of your Tuxedo configuration with the following block, load the configuration, and start the application:

```
*SERVERS
"tpadvertise.py" SRVGRP=GROUP1 SRVID=1
    REPLYQ=Y MAXGEN=2 RESTART=Y GRACE=0
    MIN=2 MAX=2
    RQADDR="tpadvertise"
```

The main difference between the SSSQ and the MSSQ configuration is the addition of the RQADDR parameter. And let's call the PING_1 service five times again:

```
echo -e "SRVCNM\tPING_1\n\n\n\n\n" | ud32
```

Unlike the first time, you should get a couple of responses saying PONG! and a couple saying UNADVERTISED called. Requests to the PING_1 service are processed by the server with SRVID 2, which does not advertise the service. Why did that happen? We experienced the limitation of the MSSQ configuration – all services using the same RQADDR parameter must advertise the same set of services. And it is completely our responsibility to enforce because Tuxedo will not. To solve the problem of unique service names, we need the second tpadvertisex function.

The tpadvertisex function takes one more argument named flags. With the default flags value of 0, it works exactly like tpadvertise. The values for flags are as follows:

- TPSECONDARYRQ: Advertise the service on a unique secondary request queue.
- TPSINGLETON: Make sure the service is unique, otherwise raise an exception.

That gives four combinations, but only both values combined give a behavior that is usable in practice.

Copy the tpadvertise.py file to tpadvertisex.py and change the line containing tpadvertise to the following:

```
        t.tpadvertisex(
            f"PING_{self.srvid}", t.TPSECONDARYRQ +
    t.TPSINGLETON
        )
```

Then, change the *SERVERS section once again to the following:

```
*SERVERS
"tpadvertisex.py" SRVGRP=GROUP1 SRVID=1
    REPLYQ=Y MAXGEN=2 RESTART=Y GRACE=0
    MIN=2 MAX=2
    RQADDR="tpadvertisex"
    SECONDARYRQ=Y
```

You should notice that we introduced one more configuration parameter, SECONDARYRQ, that adds the secondary request queue for the servers, which is off by default. Now we can try calling the PING_1 service again:

```
echo -e "SRVCNM\tPING_1\n\n\n\n\n\n" | ud32
```

This time, we will get PONG! responses, even in the MSSQ configuration.

Beyond this, if you feel curious, you can examine the server queue name, second queue name, and mapping between services and queues using the following commands:

```
echo psr | tmadmin
echo pq | tmadmin
```

However, we will continue by learning more about the services themselves.

Receiving inputs in a service

We have learned that a service is implemented using an instance method of the same name. It takes two arguments, self and data, passed to the service. In Python, data can be either a string for a STRING typed buffer, bytes for a CARRAY typed buffer, or a dictionary for an FML32 typed buffer. In addition to these two positional arguments, several named arguments can be received as well, with the most important being the following:

- name: This is the name of the service that was called. There is not much use for this in Python (unlike C), but it can be used to get the current service name without inspecting the Python function name.

- flags: May contain 0 or TPTRAN and TPNOREPLY flags. TPTRAN indicates that the service is called within a transaction. TPNOREPLY indicates that the client is not expecting a response.

If you want to examine all arguments a service receives, you can update the previous example with the following code:

```
def PING(self, data, name, flags, cd, appkey, cltid):
    t.userlog(
        f"name={name}, flags={flags}, cd={cd},"
        f" appkey={appkey}, cltid={cltid}"
    )
```

We will cover some of the additional inputs in the next chapters. It is now time to learn how a response is returned by the service.

Returning outputs from a service

Once the service has performed some computation, the response message must be returned to the caller. The function for doing that is called tpreturn. Along with the response message, it also takes arguments for indicating the success of service execution:

- rval: This is the service result for Tuxedo, and it must be one of the following three values: TPSUCCESS indicates that the service was completed successfully; TPFAIL says that the service failed, but it also marks the current transactions as rollback-only (we will learn about transactions in *Chapter 7, Distributed Transactions*); and finally, TPEXIT is similar to TPFAIL, but will cause the Tuxedo server to restart after returning the response.

- rcode: This is the service result for the application. Unfortunately, a single integer is not enough to describe the service result. I suggest using a constant, 0, for rcode, and to define errno and strerror or similar fields that contain the code and description of service execution and pass that along in the data field.

- data: This is the data service returns. Unless you have strong reasons, it is a good practice to use the input data, update it with data produced in the service, and return it.

The XATMI function tpreturn is supposed to work like the return statement in programming languages. It probably works well in C, the original implementation language. If I had a dime for every time using tpreturn was the source of memory leaks in C++ and other languages, I would have retired by now. So, tpreturn in Python only prepares the arguments for tpreturn when the function ends but does not exit the function. You have to pair it with the return statement to exit the function, like this:

```
return t.tpreturn(t.TPSUCCESS, 0, data)
```

There is another kind of tpreturn function called tpforward that is a crossbreed between tpreturn and tpcall. Let's discuss when using tpforward is beneficial next.

Optimizing tpreturn

Sometimes, the application requires a long service call chain. If calls are performed synchronously, some servers will just be waiting for responses and not processing new requests. Consider the server in the middle of the following diagram. It is waiting for the service call to complete and is not accepting any new requests nor doing any useful work.

Figure 5.1 – tpreturn call chain

Many programming languages provide a tail-call optimization that allows to perform a subroutine call as the last action of a procedure without requiring extra resources. Tuxedo has something similar. If a service call is the last action of a service, it can be performed more effectively. tpforward allows the final service call to be performed in such a way that the caller does not wait for the response but continues to process new requests. The final service in the call chain will perform a standard tpreturn function and deliver a response straight to the client, as shown in the following diagram:

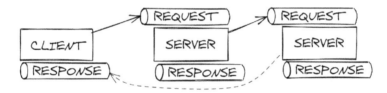

Figure 5.2 – tpforward call chain

That concludes everything you should know about servers at this point, so let's look at clients now.

Understanding client lifetime

Strictly speaking, a Tuxedo client is any program that calls services. There are two types of clients:

- A Tuxedo server can be a client if it calls other services. In this case, you should be reading the *Understanding server lifetime* section of this chapter again.

- If you are developing a Tuxedo client that is not a server, then it is no different to any other Python program except that clients call services. Nothing new here.

Let's now move on and learn about calling services.

Calling a service

The most common way of calling services is by using the tpcall function, as we did before. In addition to service name and request data, there is a third argument named flags. The possible values are as follows:

- TPSIGRSTRT means that any underlying system call that is interrupted by a signal will be retried. The short advice regarding signals is do not use them and you will not need the flag.

- TPNOBLOCK instructs to call a service in try tpcall mode. If the request queue is full, the call will return immediately instead of waiting until there is plenty of space for the request. The call will return with an exception code, TPEBLOCK.

- TPNOTIME is slightly related to the TPNOBLOCK flag. By default, the tpcall function will wait for free space in the request queue until the blocking timeout occurs. Using TPNOTIME makes the tpcall immune to this timeout at the risk of waiting for a long time. We will cover that in detail in *Chapter 7, Distributed Transactions*.

- TPNOTRAN instructs to perform the service call outside of the current transaction. Again, we will cover that in *Chapter 7, Distributed Transactions*.

The tpcall function returns a named tuple with three values and a hidden fourth value. The four values are as follows:

- rval is equal to 0 when the service returned TPSUCCESS. Or it will be equal to TPESVCFAIL when the service returned TPFAIL or TPEXIT as the first argument to the tpreturn function.

- rcode is equal to the second argument of the tpreturn function service returned.

- `data` is the service response data returned as the third and final argument of the `tpreturn` function.

- `cd` is the hidden value of *call description* that does not unpack. We will learn about this in a moment.

Here is an example of two approaches to use the *named tuple* returned by the `tpcall` function:

```
import tuxedo as t
rval, rcode, data = t.tpcall("PING_1", {})
print(rval, rcode, data)
r = t.tpcall("PING_1", {})
print(r.rval, r.rcode, r.data, r.cd)
```

You can either unpack the three values returned by the `tpcall` function or capture the *named tuple* and access each of the values individually.

So far, we have looked at the happy execution path. If anything bad happens, we will get an exception. There are many error reasons possible and you should look at Tuxedo documentation for that. One of the most common errors is calling a non-existent service, shown as follows:

```
try:
    t.tpcall("DOES_NOT_EXIST", {})
except t.XatmiException as e:
    print(e, e.code)
```

The Python module has named constants for all possible errors and you can use them to handle the exception `code` appropriately.

The `tpcall` function is what we call a synchronous service call because it waits until the service finishes and returns a result. Under the hood, `tpcall` is implemented on top of the asynchronous calls `tpacall` and `tpgetrply`. No, there is no typo. The XATMI function is called `tpgetrply`. We can implement our own `tpcall` in the following way:

```
import tuxedo as t
def my_tpcall(svc, data):
    cd = t.tpacall(svc, data)
    return t.tpgetrply(cd)
print(my_tpcall("PING_1", {}).data)
```

The tpacall function sends a request to the service but does not wait for a reply. It returns a number called a **call descriptor** and uniquely identifies the request/response pair. We can use this descriptor to receive a response for a specific request.

tpacall takes all the same inputs as the tpcall function and supports an additional flags value of TPNOREPLY that says that the client does not expect a response from the service. It is useful for sending notifications or starting asynchronous processing. The service called will also receive TPNOREPLY in service flags and may skip a costly response creation. This is how you do simple notifications:

```
t.tpacall("PAYMENT_CREATED", data, t.TPNOREPLY)
```

The tpgetrply function takes two arguments and returns the same *named tuple* as tpcall does. The first argument is the call description, and the second is flags with possible values:

- TPGETANY instructs to ignore the call descriptor in the first argument and get any response that has arrived. The real call descriptor of the received response is returned as the fourth element of the *named tuple* called cd.

- TPNOBLOCK instructs to get the response in a try tpgetrply mode. It will return a response if it is available in the response queue or raise an exception with the TPEBLOCK code.

So, what are asynchronous service calls used for? The first use is to parallelize processing. You can issue a call and continue doing some work instead of idling while waiting for a response to arrive. Here we issue a request to the PING_1 service and track the progress by checking for a response with TPNOBLOCK flags. Refer to the following code:

```
import tuxedo as t
cd = t.tpacall("PING_1", {})
while True:
    print("Waiting for a response")
    try:
        _, _, res = t.tpgetrply(cd, t.TPNOBLOCK)
        print("Got a response", res)
        break
    except t.XatmiException as e:
        if e.code == t.TPEBLOCK:
            continue
        raise
```

The second use is a scatter-gather pattern. You can send requests to multiple services at once and then wait for the responses to arrive. A classical textbook example is broadcasting quotes from multiple vendors and then selecting the best one.

Here we have a simpler example that calls *ping* services we created before and waits for all responses to arrive in any order by using the TPGETANY flag and keeping track of calls in progress:

```
import tuxedo as t
calls = []
calls.append(t.tpacall("PING_1", {}))
calls.append(t.tpacall("PING_2", {}))
while calls:
    r = t.tpgetrply(-1, t.TPGETANY)
    calls.remove(r.cd)
    print(r.data)
```

We managed to get very far in Tuxedo application development without even touching on the topic of how to *connect* to the Tuxedo application. We will look at this in the next section.

Joining the application

Before a client can access resources provided by Tuxedo servers, it must establish a *connection*. In the world of XATMI specification and Tuxedo, it is called **joining the application**. Many Tuxedo API calls do this implicitly when a function is called for the first time. The Python tuxedo module goes a step further and joins the application with non-default settings to offer more functionality by default. That should be enough for most cases and you should not think about this topic.

But there are exceptions to everything. Sometimes, an existing Tuxedo application has the AUTHSVR authentication server configured to require a login and password and you must provide it to join the application. If this is the case with you, every thread of your client program must call tpinit and tpterm functions to join to and to detach from the application. This is how the Python tuxedo module calls tpinit and tpterm behind the scenes:

```
import tuxedo as t
t.tpinit(
    usrname=None,
    cltname="tpsysadm",
```

```
    passwd=None,
    grpname=None,
    flags=t.TPMULTICONTEXTS,
)
print(t.tpcall("PING_1", {}).data)
t.tpterm()
```

If you require a login and password, supply them as usrname and passwd arguments.

Tuxedo servers can also act as clients and access resources provided by other servers. Even if the application requires a login and password, the server will be able to access it with no changes. The only exceptions are threads created programmatically using Python's threading module or other libraries. Since Tuxedo does not know about such threads, each thread must call the tpappthrinit and tpappthrterm functions to join the application. This is what the tuxedo module does behind the scenes:

```
import tuxedo as t
t.tpappthrinit(
    usrname=None,
    cltname="tpsysadm",
    passwd=None,
    grpname=None,
    flags=t.TPMULTICONTEXTS,
)
print(t.tpcall("PING_1", {}).data)
t.tpappthrterm()
```

If your application needs a login and password, a call to tpappthrinit with usrname and passwd arguments must be the first thing you do in a newly created thread. Once it is done, the thread can proceed to call services and interact with the Tuxedo application.

As in the case of clients, change the arguments according to the expectations of the application.

Summary

In this chapter, we developed multiple clients and servers. I hope it connected various bits and pieces of Tuxedo architecture, typed buffers, message passing, and clients and servers. We witnessed how SSSQ and MSSQ configuration impacts the Tuxedo application and how to avoid its quirks. We learned how to build service pipelines, how to send notifications, and how to do parallel processing using primitives offered by Tuxedo. You also had an insight into what the Python library does behind the scenes to simplify your day.

After this and the previous chapters, you should have enough theoretical knowledge to develop Tuxedo applications. We can now continue to more practical topics and start with administration and monitoring of Tuxedo applications using Python code in the next chapter.

Questions

1. What function is used to advertise a service on the secondary queue?

2. Are `tpsvrthrinit` and `tpsvrthrdone` called for single-threaded servers?

3. What flag tells `tpacall` to not expect a response?

4. What flag tells `tpgetrply` to get the first response available?

5. What flag tells a function to *try* the call?

Further reading

-   Tuxedo documentation: `https://docs.oracle.com/cd/E53645_01/tuxedo/docs12cr2/rf3c/rf3c.html`

6
Administering the Application Using MIBs

So far, in this book, we have been creating a Tuxedo configuration file and performing a full cycle that involves loading the configuration, starting the application, and stopping the application. However, Tuxedo also supports changing the configuration of running applications through the **Management Information Base** (**MIB**). By using MIBs, applications can be upgraded without any downtime, and the administrator can gain more insight into what and how the application is performing.

In this chapter, we will cover the following topics:

- Introducing MIB
- Developing multiple versions of the application
- Using `tpadmcall` to create `TUXCONFIG`
- Upgrading the servers
- Reconfiguring the application
- Monitoring the application

By the end of this chapter, you will understand how to create the initial configuration programmatically. You will know how to perform basic upgrades and even complicated upgrades with configuration changes without any downtime. Finally, you will understand how to retrieve more information about the Tuxedo application than the standard tools provide.

Technical requirements

All of the code for this chapter is available at `https://github.com/PacktPublishing/Modernizing-Oracle-Tuxedo-Applications-with-Python/tree/main/Chapter06`.

The Code in Action video for the chapter can be found at `https://bit.ly/3vE204T`.

Introducing MIB

MIB is a unified interface for the administration of Tuxedo applications. The interface uses `FML32` typed buffers and fields that are defined in files under the `$TUXDIR/udataobj` directory of a Tuxedo installation. There are multiple MIBs for different Tuxedo components, such as Domains, EventBroker, and Workstation. However, each of these components is beyond the scope of this book. Here, we will keep our focus on `TM_MIB`, which provides information about the Tuxedo core.

The most common way to interact with the MIB is by calling the `.TMIB` service provided by the `BBL` server. You can do this without writing any code by using the `ud32` utility. Note that we have used this in our examples in the previous chapters. The one obvious limitation of using the `.TMIB` service is that the `BBL` server must be kept running.

The second way to interact with the MIB is by using the `tpadmcall` API function, which works even when the application is not running or has not been configured at all. When the application is running, `tpadmcall` can be used in read-only mode only, and no modifications can be performed. There are more nuances to this, but this simplified view is good enough for most situations.

All MIB requests contain the following common fields:

- `TA_OPERATION` is `SET`, to modify the configuration, or `GET` and `GETNEXT`, to retrieve the information.

- `TA_CLASS` identifies the component that we want to access. Each component then has additional input and output fields, as described in the documentation.

- `TA_STATE` represents the status of the record. By changing the values of this field, you can create a new record, start and stop a server, or destroy a record.

- `TA_FLAGS` refers to the flags for the request. The most important value is `MIB_LOCAL`, which requests to include additional information in the response.

- `TA_CURSOR` is used for pagination along with the `TA_OPERATION` value of `GETNEXT`.

- `TA_FILTER` contains a list of field identifiers to return; otherwise, everything is returned. The pagination of information retrieval is not based on the number of records but on the size of the output. Using `TA_FILTER` is useful for retrieving all necessary information in as few requests as possible. The values in this list are field identifiers, not field names. To get the field identifier from the field name, use the `Fldid32` function.

The important output fields that you should know are listed here:

- `TA_ERROR`: This is a return code characterizing the result of the operation. Negative values indicate an error; 0 and positive values indicate no updates made, all updates made, or partial updates made. For most cases, it is sufficient to check whether `TA_ERROR` is less than 0.

- `TA_STATUS`: This contains a more detailed description of the error.

- `TA_BADFLD`: This field contains a list of field identifiers (not names) causing the error. To get the name of the field, use the `Fname32` function that we learned about in *Chapter 4, Understanding Typed Buffers*.

- `TA_MORE`: This field, which is for information retrieval, indicates that there is more data to retrieve. In this scenario, the `TA_CURSOR` field will be present with a value to retrieve the next page.

> **Tip**
>
> As you can see, even Tuxedo does not rely only on the application return code, `rcode`. It uses a pair of fields—one for the error code and the other for the error description. Your applications should do the same. Not only does this provide more feedback to the client, but it also makes it easier to either preserve or pass along the data.

To better understand how MIB works, let's try it out in practice.

Developing multiple versions of an application

In this section, we will develop two versions of an application that return slightly different outputs and perform upgrades and downgrades between the versions. To do this, we will need a client program to generate some load and to ensure upgrades can be performed without any downtime:

1. First, create a client program, called `load.py`, with the following content:

```python
import time
import tuxedo as t
seconds = int(time.time())
status = []
while True:
    status.extend(t.tpcall("PING", {}).data["TA_STATUS"])
    if seconds != int(time.time()):
        print(
            "{} | cps={} | status={}".format(
                time.ctime(seconds), len(status),
set(status)
            )
        )
        seconds = int(time.time())
        status = []
    time.sleep(0.01)
```

We call the `PING` service in a loop and gather the `TA_STATUS` response values. At each second, we print the current time, the number of responses received, and the unique response values. We also suspend execution for 0.01 seconds between calls so that we do not overload the application.

2. Next, we need two versions of the server. The first one is called `ping1.py` and has the following content:

```python
#!/usr/bin/env python3
import sys
import tuxedo as t
class Server:
    def tpsvrinit(self, argv):
        t.tpadvertise("PING")
        return 0
    def PING(self, req):
        req["TA_STATUS"] = "v1"
        return t.tpreturn(t.TPSUCCESS, 0, req)
t.run(Server(), sys.argv)
```

We advertise the PING service, which returns TA_STATUS with v1 as the value.

3. The second version, called `ping2.py`, is the same except it returns v2 as the value of TA_STATUS:

```python
#!/usr/bin/env python3
import sys
import tuxedo as t
class Server:
    def tpsvrinit(self, argv):
        t.tpadvertise("PING")
        return 0
    def PING(self, req):
        req["TA_STATUS"] = "v2"
        return t.tpreturn(t.TPSUCCESS, 0, req)
t.run(Server(), sys.argv)
```

Remember to make your files executable by running the following:

```
chmod +x ping1.py ping2.py
```

Additionally, make sure you have a copy of `ping1.py`, called `ping.py`:

```
cp ping1.py ping.py
```

Now we are ready to create a configuration file and run the application. We can do that by creating a ubbconfig file in a text editor; however, this is pretty boring since we could be writing code instead!

Using tpadmcall to create TUXCONFIG

By using tpadmcall, we can skip the textual configuration file and create the binary configuration file directly. To avoid cryptic error messages, ensure you have no other application running and that the file in the $TUXCONFIG environment variable does not exist. To shut down the application and to remove the file in the $TUXCONFIG environment variable, run the following commands:

```
tmshutdown -y
rm -f $TUXCONFIG
```

The configuration files we created so far consist of RESOURCES, MACHINES, GROUPS, and SERVICES sections. The MIB operates on the same sections but calls them *classes*.

We start by creating the T_DOMAIN record, which is the same as the RESOURCES section, the textual representation, and by specifying the same parameter values. The main difference when compared to the configuration file is the TA_ prefix for the parameter name, as shown in the following code snippet:

```
import tuxedo as t
t.tpadmcall({
        "TA_CLASS": "T_DOMAIN",
        "TA_OPERATION": "SET",
        "TA_STATE": "NEW",
        "TA_MASTER": "tuxapp",
        "TA_MODEL": "SHM",
        "TA_IPCKEY": 32769,
})
```

By creating a T_DOMAIN record, we have also implicitly created a T_MACHINE record, which is the same as the MACHINES section in the textual representation. You can verify that by unloading the configuration using the tmunloadcf command. The default values filled in for the machine are good enough for our application.

Next, we create a T_GROUP record for the GROUPS section:

```
t.tpadmcall({
        "TA_CLASS": "T_GROUP",
        "TA_OPERATION": "SET",
        "TA_STATE": "NEW",
        "TA_SRVGRP": "GROUP1",
        "TA_GRPNO": 1,
        "TA_LMID": "tuxapp",
})
```

Now, let's create a T_SERVER record for the SERVERS section:

```
t.tpadmcall({
        "TA_CLASS": "T_SERVER",
        "TA_OPERATION": "SET",
        "TA_STATE": "NEW",
        "TA_SRVGRP": "GROUP1",
        "TA_SERVERNAME": "ping.py",
        "TA_SRVID": 1,
        "TA_MIN": 2,
        "TA_MAX": 2,
        "TA_REPLYQ": "Y",
        "TA_MAXGEN": 2,
        "TA_RESTART": "Y",
        "TA_GRACE": 0,
        "TA_RQADDR": "ping",
})
```

We use the ping.py binary that contains the v1 version of the PING service. And we start two instances of the service in the MSSQ configuration.

And that is all! Once you have executed all of the preceding Python code, you will have the binary Tuxedo configuration file, tuxconfig, ready. We can skip the loading of the textual configuration file since we don't have one and start the application straight away:

```
tmboot -y
```

The application should start and we are ready to do an upgrade. I do not want to scare you with a complicated upgrade, so we will do a simple upgrade first.

Upgrading the servers

The most common upgrade is simply changing the executable files of the Tuxedo servers. Because we want to observe how our application behaves, we need a second terminal for this exercise. If you are using your own Linux machine, just start a new shell. If you followed the book and used Docker, then you must find the current container name by using this command:

```
cat /etc/hostname
```

In my case, it outputs `15c365dcb562`. Then, you can start a new shell in the container by using the container name that you have obtained:

```
docker exec -it 15c365dcb562 bash
```

If you are using Microsoft Windows, you might need to add `winpty` to the front of the command. You will also need to change the user to Oracle:

```
su - oracle
```

There are still some environment variables that need to be set up. Assuming that you are in the folder where your code resides, you can use the following commands:

```
source /home/oracle/tuxhome/tuxedo12.2.2.0.0/tux.env
export TUXCONFIG=`pwd`/tuxconfig
export FLDTBLDIR32=$TUXDIR/udataobj:`pwd`
export FIELDTBLS32=Usysfl32,tpadm,example
```

This should be enough to run the client program and start generating some load:

```
python3 load.py
```

We will observe the output of this command while we upgrade our application. Leave it running and return to the first terminal.

For the upgrade, we must install a new version of the server. My experience has taught me that replacing the binary file sometimes leads to crashes. So, instead of replacing the file, it is better to rename the file because it keeps the file content in the same physical location on the disk. You can now create a new file with the old name for Tuxedo to pick up:

```
mv ping.py ping.old
cp ping2.py ping.py
```

Now we can restart each instance of the server. There should be no downtime for our application because the second instance of the server will continue handling requests while the first is being restarted:

```
tmshutdown -i 1
tmboot -i 1
tmshutdown -i 2
tmboot -i 2
```

In the second terminal running load.py, you will observe the following:

- There is no change in the number of calls per second during the stopping and starting of server instances.

- After restarting the first instance of the server, there will be mixed responses from both the first and second versions.

You can also go back to the first version of the Tuxedo server if the new version does not work as expected by running the following command:

```
mv ping.py ping.old && cp ping1.py ping.py
tmshutdown -i 1 && tmboot -i 1 && tmshutdown -i 2 && tmboot -i
2
```

You should see, in the second terminal, that the application is running version v1 again.

> **Tip**
> Servers in Tuxedo are uniquely identified by a combination of a group identifier and a server identifier within that group. However, it is a good practice to keep the server identifier unique within the application, not just the group. It greatly simplifies many tasks, such as starting and stopping using just the server identifier.

If you want to spend more time on this topic, here are some ideas for experiments that you can carry out:

- Remove the time.sleep line from load.py to see whether there is a difference in the number of calls per second when the application is under maximum load.

- We experimented with the MSSQ configuration. Instead, you can try the SSSQ configuration and see if, and how, load balancing changes the outcome.

Binary upgrades and downgrades are simple, but often, a reconfiguration of the application is required.

Reconfiguring the application

Up until now, whenever we needed to change the configuration of the application, we stopped the application, loaded the new configuration, and started it again. For a real application with hundreds of servers, this requires significant downtime.

In this section, we will replace two instances of the ping.py server with five instances of the ping2.py server. To make it more interesting, we will also create a new group for them because there are times when you also have to modify the group configuration. And finally, we will remove the old instances of these servers.

To change the configuration while the application is running, we have to call the .TMIB service. In addition to changing the configuration, it can also start and stop Tuxedo servers by changing the value of TA_STATE from INAtive to ACTive and back. Keep python3 load.py running and watch how the application behaves.

First, we create a new group, called GROUP2, and start it:

```
import tuxedo as t
t.tpcall(".TMIB", {
        "TA_CLASS": "T_GROUP",
        "TA_OPERATION": "SET",
        "TA_STATE": "NEW",
        "TA_SRVGRP": "GROUP2",
        "TA_GRPNO": 2,
        "TA_LMID": "tuxapp",
})
t.tpcall(".TMIB", {
        "TA_CLASS": "T_GROUP",
        "TA_OPERATION": "SET",
        "TA_STATE": "ACT",
        "TA_SRVGRP": "GROUP2",
})
```

Then, we can create the configuration for five `ping2.py` servers. Once the configuration is created, we can start the servers one by one:

```
t.tpcall(".TMIB", {
        "TA_CLASS": "T_SERVER",
        "TA_OPERATION": "SET",
        "TA_STATE": "NEW",
        "TA_SRVGRP": "GROUP2",
        "TA_SERVERNAME": "ping2.py",
        "TA_SRVID": 10,
        "TA_MIN": 5,
        "TA_MAX": 5,
        "TA_REPLYQ": "Y",
        "TA_MAXGEN": 2,
        "TA_RESTART": "Y",
        "TA_GRACE": 0,
        "TA_RQADDR": "ping2",
})
for srvid in range(10, 10 + 5):
    t.tpcall(".TMIB", {
            "TA_CLASS": "T_SERVER",
            "TA_OPERATION": "SET",
            "TA_STATE": "ACT",
            "TA_SRVGRP": "GROUP2",
            "TA_SRVID": srvid,
    })
```

To remove the old servers, we perform the same steps in reverse order. First, we have to stop the servers and delete the configuration:

```
for srvid in range(1, 1 + 2):
    t.tpcall(".TMIB", {
            "TA_CLASS": "T_SERVER",
            "TA_OPERATION": "SET",
            "TA_STATE": "INA",
            "TA_SRVGRP": "GROUP1",
            "TA_SRVID": srvid,
    })
```

```
t.tpcall(".TMIB", {
        "TA_CLASS": "T_SERVER",
        "TA_OPERATION": "SET",
        "TA_STATE": "INV",
        "TA_SRVGRP": "GROUP1",
        "TA_SRVID": 1,
})
```

You can then safely remove GROUP1 as it is no longer in use. To do so, run the following:

```
t.tpcall(".TMIB", {
        "TA_CLASS": "T_GROUP",
        "TA_OPERATION": "SET",
        "TA_STATE": "INA",
        "TA_SRVGRP": "GROUP1",
})
t.tpcall(".TMIB", {
        "TA_CLASS": "T_GROUP",
        "TA_OPERATION": "SET",
        "TA_STATE": "INV",
        "TA_SRVGRP": "GROUP1",
})
```

The application now contains five instances of the ping2.py server and none of the ping.py server. You can verify this by using the commands that we learned earlier:

```
echo psr | tmadmin
```

This will show five servers. The changes carried out by our Python script are permanent. You can stop the application and start it up again, and it will contain the same servers. All changes have been stored in the Tuxedo binary configuration file, which is found in the TUXCONFIG environment variable. You can verify this by exporting the textual representation of the configuration by using the tmunloadcf command.

So far, we have used MIB to access and modify the same information that is present in the configuration files. However, there is more information in MIB than that.

Monitoring the application

In addition to the configuration, MIB also contains runtime information and statistics about the application. We already viewed some of this when we used the tmadmin command in previous chapters.

To get an overview of the system statistics, we will use the tpadmcall function. Since tpadmcall can work both with a running and stopped application, it assumes a stopped application by default and does not attempt to connect it. However, we now have a running application and we must connect it explicitly. We will use the tpsysop client name to obtain read-only access to the system. We will also set a special flags value, called MIB_LOCAL, which instructs the application to return local information containing statistics:

```
import tuxedo as t
t.tpinit(cltname="tpsysop")
machine = t.tpadmcall({
        "TA_CLASS": "T_MACHINE",
        "TA_OPERATION": "GET",
        "TA_FLAGS": t.MIB_LOCAL,
}).data
```

We have obtained several interesting fields now, and all of them are described in the Tuxedo documentation. For me, the most important of these include the current timestamp, the total number of service calls made, and the transaction statistics. Service calls show how busy the application is and the transaction number gives us an insight into the error rate, and it is a good metric to keep an eye on:

```
print("Timestamp:", machine["TA_CURTIME"][0])
print("Service calls:", machine["TA_NUMREQ"][0])
print("Transactions started:", machine["TA_NUMTRAN"][0])
print("Transactions aborted:", machine["TA_NUMTRANABT"][0])
print("Transactions committed:", machine["TA_NUMTRANCMT"][0])
```

The nice thing about using the `tpadmcall` function is that it will not change the number of service calls and the application will not be impacted by measuring it. The transaction numbers will be 0 at the moment, but that will change in *Chapter 7, Distributed Transactions.*

Monitoring queues

We have already learned how to monitor queues and the number of service requests completed by servers by using the `pq`, `psr`, and `psc` commands in `tmadmin`. However, the MIB interface contains even more information. While using the `tpadmcall` function will work on our simple application, real applications have more information over multiple pages, and the only way to retrieve all of these pages is using by the `.TMIB` service call instead.

First, let's wrap the `.TMIB` service call in a function that retrieves all of the pages of information:

```python
import os
import tuxedo as t
def getall(req):
    res = t.tpcall(".TMIB", req).data
    out = res
    while res["TA_MORE"][0] > 0:
        req["TA_OPERATION"] = "GETNEXT"
        req["TA_CURSOR"] = res["TA_CURSOR"]
        res = t.tpcall(".TMIB", req).data
        for k, v in res.items():
            out[k].extend(v)
    return out
```

We will need to retrieve three classes of information from the MIB. The `T_SERVER` records about servers might be familiar to you as we have used them in earlier examples. The `T_CLIENT` and `T_MSG` records about clients and message queues are purely runtime information and cannot be created or modified. Now, we can use our `getall` function to retrieve all of the pieces of information, as shown here:

```python
server = getall({
        "TA_CLASS": "T_SERVER",
        "TA_OPERATION": "GET",
        "TA_FLAGS": t.MIB_LOCAL,
```

```
})
client = getall({
        "TA_CLASS": "T_CLIENT",
        "TA_OPERATION": "GET",
        "TA_FLAGS": t.MIB_LOCAL,
})
msg = getall({
        "TA_CLASS": "T_MSG",
        "TA_OPERATION": "GET",
        "TA_FLAGS": t.MIB_LOCAL,
})
```

To obtain a nicer output, we will resolve each PID to the Tuxedo server name or client. Client processes do not have names, so, in practice, it is better to rely on the output of Unix command ps. We will try to keep this as simple as possible:

```
processes = {}
for i in range(len(server["TA_PID"])):
    processes[server["TA_PID"][i]] = os.path.basename(
        server["TA_SERVERNAME"][i]
    )
for pid in client["TA_PID"]:
    processes[pid] = "client#{}".format(pid)
```

Finally, we can process the System V message queue statistics. For each queue, we receive the number of messages and bytes in the queue, the PID of the last reader and writer, and the timestamps. With this information, we can display which processes were communicated through the queues and, therefore, get a sense of message flows and queueing within the Tuxedo application:

```
for i in range(len(msg["TA_CURTIME"])):
    print(
        "{} : {} -> [{} msg {}/{} bytes] -> {}".format(
            max(msg["TA_MSG_RTIME"][i], msg["TA_MSG_STIME"]
[i]),
                processes.get(
                    msg["TA_MSG_LRPID"][i], msg["TA_MSG_LRPID"][i]
                ),
                msg["TA_MSG_QNUM"][i],
```

```
        msg["TA_MSG_CBYTES"][i],
        msg["TA_MSG_QBYTES"][i],
        processes.get(
            msg["TA_MSG_LSPID"][i], msg["TA_MSG_LSPID"][i]
        ),
    )
)
```

The best part about this is that Python does not come with a library that provides information about IPC queues. The MIB returns all of it, not only for the request queues but for the response queues as well.

> **Tip**
> One important, if not the most important, task of Tuxedo application administration is to ensure that the queues never fill up. Combining queue information with client and server information offers a better insight into what the application is doing and helps us to investigate any incidents.

Summary

In this chapter, we learned how to use MIB to configure both stopped and running Tuxedo applications. We performed an upgrade and a reconfiguration of the running application without creating any downtime or causing any visible impact. We also learned how to retrieve monitoring information from Tuxedo.

After reading this chapter, you should be able to perform many administration tasks with your Tuxedo application. You also recreated one of my favorite client, queue, and server monitoring scripts. Since Tuxedo itself is just a tool to develop applications, this might not be enough. However, you now have a good foundation to build on by exploring the huge Tuxedo MIB documentation and enhancing it with features that are specific to your application.

In the next chapter, you will learn about distributing transactions and how they work in Tuxedo.

Questions

1. Which of `tpadmcall` and `tpcall(.TMIB)` can be used to create the initial Tuxedo configuration?

2. Which of `tpadmcall` and `tpcall(.TMIB)` can be used to reconfigure a running Tuxedo application?

3. Is MIB pagination based on the number of records or the size of the result?

4. What is the flag value that is needed to retrieve statistical information?

Further reading

MIB and the `TM_MIB` documentation can be found at `https://docs.oracle.com/cd/E53645_01/tuxedo/docs12cr2/rf5/rf5.html`.

7
Distributed Transactions

Tuxedo stands for **Transactions for Unix, Extended for Distributed Operations** and it is described as a transaction processing system or a transaction monitor. It should be no surprise that transactions are an important part of Tuxedo. Most Tuxedo applications use transactions, and understanding how they work is essential for developing Tuxedo applications.

In this chapter, we will cover the following topics:

- Configuring Tuxedo for transactions
- Managing transactions
- Understanding timeouts

By the end of this chapter, you will know how to enable transaction support in Tuxedo. You will create a transaction and configure transaction manager servers. You will learn different ways of creating transactions by configuration and by writing code. You will learn how to detect whether your code executes within a transaction and how to escape the transaction if needed. Finally, you will learn how to use transaction timeouts, blocking timeouts, and avoiding typical timeout problems.

Technical requirements

All the code for this chapter is available at `https://github.com/ PacktPublishing/Modernizing-Oracle-Tuxedo-Applications-with- Python/tree/main/Chapter07`.

The Code in Action video for the chapter can be found at `https://bit.ly/3qXbxAL`.

Configuring Tuxedo for transactions

A transaction is a unit of work that has **ACID** properties—**Atomic**, **Consistent**, **Isolated**, and **Durable**. But Tuxedo takes transactions a step further and supports distributed transactions that provide ACID properties for two or more databases from different vendors, queueing systems, and filesystems. Tuxedo was also the basis for the X/Open XA specification, which is the de facto standard for distributed transactions.

Transactions are infectious in Tuxedo. Once a transaction is started in Tuxedo, all the following service calls become a part of the transaction unless otherwise specified. If any of the services participating in the transaction fails, the whole transaction is marked as bad and must be aborted.

Although distributed transactions and two-phase commit do have problems and limitations, they work well in Tuxedo. Distributed transactions are not as trendy as **Basically Available**, **Soft state**, **Eventual consistency** (**BASE**) properties, the Saga pattern, and compensating transactions. But if you are using Tuxedo, just give distributed transactions a try. Tuxedo tries hard to apply many optimizations to reduce distributed transactions to a simple transaction by merging Oracle database transactions, detecting read-only transaction branches, and sorting resources involved in the transaction.

Transactions can be started by both clients and servers. This can be done programmatically or by using Tuxedo configuration. The easiest way, however, is to start a transaction by using the `ud32` tool. It supports a `-t` command-line argument that instructs you to start a transaction with a specified transaction timeout. We can use it on any application we have developed so far. Let's call the `.TMIB` service within a transaction:

```
echo -e "SRVCNM\t.TMIB\n" | ud32 -t 10
```

You will get an error saying that Tuxedo failed to start a transaction. That is because we did not enable transaction support in Tuxedo.

To use transactions, we must first create a transaction log. This is a special file where Tuxedo stores information about the state of transaction branches. To be very specific, it does so only between the first and the second phase of the commit. The transaction log should not become the bottleneck but it is still a good practice to put it on fast disk drives. The latest versions of Tuxedo since 12.1.3 support storing transaction logs in the Oracle database, but you will have to learn about that from the Tuxedo documentation. To create a transaction log, we must create the Tuxedo configuration `ubbconfig` file with the following content:

```
*RESOURCES
MASTER tuxapp
MODEL SHM
IPCKEY 32769
*MACHINES
"15c365dcb562" LMID=tuxapp
    TUXCONFIG="/home/oracle/code/07/tuxconfig"
    TUXDIR="/home/oracle/tuxhome/tuxedo12.2.2.0.0"
    APPDIR="/home/oracle/code/07"
    TLOGDEVICE="/home/oracle/code/07/tlog"
    TLOGSIZE=100
*GROUPS
GROUP1 LMID=tuxapp GRPNO=1 TMSNAME=TMS TMSCOUNT=2
*SERVERS
"ping.py" SRVGRP=GROUP1 SRVID=1
    REPLYQ=Y MAXGEN=2 RESTART=Y GRACE=0
    MIN=2 MAX=2
    RQADDR="ping"
```

Here are the most significant changes compared to the configuration we have used so far:

- `TLOGDEVICE`: This is the absolute path to the transaction log, which we will create next.

- `TLOGSIZE`: This is the maximum number of transactions that can be logged at a given time. The default value is `100`.

- TMSNAME: This is the name of the transaction manager server. It is a special server that knows how to commit and roll back transactions for all servers belonging to the group. The TMS value is a special dummy transaction manager that allows any server that does not need or use transactions to participate in a global transaction. We will learn about other transaction managers in *Chapter 8, Using Tuxedo Message Queue*, and *Chapter 9, Working with Oracle Database*.

- TMSCOUNT: This is the number of transaction managers that will be started. 2 is the minimum value and we use it because the default value of 3 will make the application startup slightly longer.

After creating the ubbconfig file, we can load it:

```
tmloadcf -y ubbconfig
```

Now is the time to create the transaction log. To do this, we use the tmadmin utility and command, which takes the full path to the transaction log and the size of the log in blocks. Each transaction takes up one block. In ubbconfig, we specified that the transaction log size will be 100 blocks/transactions. But Tuxedo needs some blocks for housekeeping, and the size needed is somewhere between 100 and 200. We will allocate excess space just to be safe:

```
echo "crdl -z `pwd`/tlog -b 200" | tmadmin
```

The previous command created *a device*, not the transaction log itself. Think of this *device* as a file of a fixed size where write operations cannot fail due to a lack of disk space. Now we create the real transaction log inside the device by specifying the machine identifier of our application:

```
echo "crlog -m tuxapp" | tmadmin
```

Now the configuration is done and we lack only a Tuxedo server to run. We will start with the following ping.py server:

```
#!/usr/bin/env python3
import sys
import tuxedo as t
class Server:
    def tpsvrinit(self, argv):
        t.tpadvertise("PING")
        return 0
    def PING(self, req, flags):
```

```
        if flags & t.TPTRAN:
            t.userlog("PING in a transaction")
        else:
            t.userlog("PING outside a transaction")
        req["TA_STATUS"] = "v1"
        return t.tpreturn(t.TPSUCCESS, 0, req)
t.run(Server(), sys.argv)
```

This returns a version number in the TA_STATUS field and checks the service flags for the TPTRAN bit to see whether it is called within a transaction. Do not forget to make the file executable; then, we can start the application:

```
tmboot -y
```

You will notice that the Tuxedo starts two instances of the TMS server before starting our ping.py server. Try running the following command:

```
echo psc | tmadmin
```

You will see that the server identifiers for TMS servers are outside the SRVID range allowed for regular Tuxedo servers. They advertise a TMS service.

We can test whether transactions work. Try calling the PING service within a transaction:

```
echo -e "SRVCNM\tPING\n" | ud32 -t 5
```

Now check ULOG and see whether the PING service was called in or outside a transaction. It will be in a transaction and the ULOG entry will be prefixed with a strange hexadecimal transaction identifier such as gtrid x0 x6011d7fd x4. You can also call the service outside the transaction:

```
echo -e "SRVCNM\tPING\n" | ud32
```

It will behave just like the services we developed in previous chapters; enabling transactions did not change the non-transactional service calls.

At the beginning of the chapter, we tested transaction support by calling the .TMIB service. That was a slightly dishonest example because the .TMIB service and the BBL server are special; they do not have a transaction manager server and do not support transaction intentionally. But now, let's learn how to manage a transaction in the application code.

Managing transactions

Transactions in Tuxedo can be started by using the `tpbegin` function call with the transaction timeout in seconds as the parameter. The transaction must be either committed by using the `tpcommit` function or aborted by using the `tpabort` function.

> **Tip**
>
> I recommend creating a Python Context Manager for transactions to ensure the transaction is always completed by committing or aborting it.

Here is an example of how to create a transaction and call service within a transaction:

```
import tuxedo as t
t.tpbegin(3)
t.tpcall("PING", {})
t.tpcommit()
```

ULOG will show that the service was called in a transaction although we did nothing special for the service invocation. Transactions in Tuxedo are infectious: once the code executes in a transaction, all service calls are performed in the same transaction. How can you know if the code executes in a transaction? One way of knowing is to inspect the `flags` service that it receives in parameters as we did in the PING service. The other bullet-proof method is to call the `tpgetlev` function. We can sprinkle our code with it and check the result:

```
import tuxedo as t
print("In transaction?", t.tpgetlev())
t.tpbegin(3)
print("In transaction?", t.tpgetlev())
t.tpcall("PING", {})
print("In transaction?", t.tpgetlev())
t.tpcommit()
print("In transaction?", t.tpgetlev())
```

You will see that there is no active transaction before calling `tpbegin` and after calling `tpcommit`.

Sometimes you have to call a service that cannot participate in a transaction due to a lack of a transaction management server. Or maybe you want to store some evidence even when the main transaction is canceled. To achieve that, you have to prevent the spreading of the transactions, and that requires some extra code. The simplest way is to add the TPNOTRAN flag to the `tpcall` or `tpacall` function:

```
import tuxedo as t
t.tpbegin(3)
t.tpcall("PING", {}, t.TPNOTRAN)
t.tpcommit()
```

ULOG will show that the service was called outside the transaction. There is a second way to do so, by suspending the current transaction and resuming it later. The functions are `tpsuspend`, which suspends transaction and returns a transaction identifier, and `tpresume`, which resumes the transaction given in the parameters:

```
import tuxedo as t
t.tpbegin(3)
trxid = t.tpsuspend()
t.tpcall("PING", {})
t.tpresume(trxid)
t.tpcommit()
```

That will give you the same result as the previous example but without the TPNOTRAN parameter for the `tpcall` function. However, your first choice to escape transactions should be to use TPNOTRAN. The main use for suspending a transaction is when you want a new transaction in the middle of another transaction like this:

```
import tuxedo as t
t.tpbegin(3)
trxid = t.tpsuspend()
t.tpbegin(1)
t.tpcall("PING", {})
t.tpcommit()
t.tpresume(trxid)
t.tpcommit()
```

This section would not be complete without mentioning that transactions can also be managed in the ubbconfig configuration file by using the AUTOTRAN feature:

```
*RESOURCES
MASTER tuxapp
MODEL SHM
IPCKEY 32769
*MACHINES
"15c365dcb562" LMID=tuxapp
     TUXCONFIG="/home/oracle/code/07/tuxconfig"
     TUXDIR="/home/oracle/tuxhome/tuxedo12.2.2.0.0"
     APPDIR="/home/oracle/code/07"
     TLOGDEVICE="/home/oracle/code/07/tlog"
     TLOGSIZE=100
*GROUPS
GROUP1 LMID=tuxapp GRPNO=1 TMSNAME=TMS TMSCOUNT=2
*SERVERS
"ping.py" SRVGRP=GROUP1 SRVID=1
     REPLYQ=Y MAXGEN=2 RESTART=Y GRACE=0
     MIN=2 MAX=2
     RQADDR="ping"
*SERVICES
PING AUTOTRAN=Y TRANTIME=3
```

We have introduced the SERVICES section, which contains configuration for services:

- The first value is the Tuxedo service name, PING in our example. It must be up to *127* characters long, the same as the limitation of the service name length.

- AUTOTRAN: This specifies whether the transaction should be started if the service is not already called in a transaction. The default value is N and we can turn this feature on by using the value Y.

- TRANTIME: This is the transaction timeout in seconds in case the transaction is started by Tuxedo. The default value is 30 seconds.

When we start the application using this configuration, no matter how we call the service, it will execute within the transaction. Even if we call the service with TPNOTRAN flags, Tuxedo will start a new transaction for the service and commit it if the service returns with TPSUCCESS or abort the transaction if the service returns TPFAIL or TPEXIT. The only difference you may notice in practice is different transaction timeouts when the service was called as part of the transaction, or that Tuxedo started a new transaction for it.

This brings us nicely to the next topic of transaction timeouts and timeouts in general.

Understanding timeouts

What is the point of using transactions in this chapter if we are not using any databases? Besides making the Tuxedo transaction API easier to explain, its most useful feature is timeouts. Timeouts allow enforcing time constraints for computation.

There are two kinds of timeouts in Tuxedo, and while they are different, their implementation and behavior are similar. You already know about transaction timeouts. The second kind is called *blocking* timeouts. The name comes from Unix jargon, where *blocking* describes a system call that may wait indefinitely until it completes or fails with an error; this is as opposed to the *non-blocking* mode, where the system call completes or fails immediately.

The main thing to remember about timeouts in Tuxedo is that they apply to the clients, not the servers. The client will receive the timeout error but the server will continue working for hours with no clue that timeout occurred. The only way for the server to know that it exceeded the timeout is if database sessions have appropriate timeouts and a database operation returns an error or the server becomes a client by calling some other service in a transaction. To understand this better, we will have to develop an application.

We have to start with the configuration file once again:

```
*RESOURCES
MASTER tuxapp
MODEL SHM
IPCKEY 32769
SCANUNIT 5
BLOCKTIME 3
*MACHINES
"15c365dcb562" LMID=tuxapp
    TUXCONFIG="/home/oracle/code/07/tuxconfig"
```

```
        TUXDIR="/home/oracle/tuxhome/tuxedo12.2.2.0.0"
        APPDIR="/home/oracle/code/07"
        TLOGDEVICE="/home/oracle/code/07/tlog"
        TLOGSIZE=100
*GROUPS
GROUP1 LMID=tuxapp GRPNO=1 TMSNAME=TMS TMSCOUNT=2
*SERVERS
"ping.py" SRVGRP=GROUP1 SRVID=1
        REPLYQ=Y MAXGEN=2 RESTART=Y GRACE=0
        MIN=2 MAX=2
        RQADDR="ping"
```

The important changes compared to the previous configuration are as follows:

- SCANUNIT: This is the number of seconds between periodic checks performed by BBL and must be a multiple of *2* or *5*. The latest versions of Tuxedo support a SCANUNIT value of milliseconds by using the MS suffix.

- BLOCKTIME: This is a multiplier for SCANUNIT to get a blocking timeout. *5 * 3* gives a *15*-second blocking timeout.

Due to the way Tuxedo implements all timeouts by having a periodic check, the timeout value is not exact. Here, we have configured a timeout of *15* seconds. If Tuxedo performs a periodic check in the *14th* second, we will not get the timeout in *15* seconds but in *19* seconds when Tuxedo performs the next periodic check.

Next, let's enhance our ping.py with another service that will simulate a long-running process:

```
#!/usr/bin/env python3
import sys
import time
import tuxedo as t
class Server:
    def tpsvrinit(self, argv):
        t.tpadvertise("PING")
        t.tpadvertise("SLEEP")
        return 0
    def PING(self, req, flags):
        if flags & t.TPTRAN:
```

```
            t.userlog("PING in a transaction")
        else:
            t.userlog("PING outside a transaction")
        req["TA_STATUS"] = "v1"
        return t.tpreturn(t.TPSUCCESS, 0, req)
    def SLEEP(self, req, flags):
        for _ in range(10):
            t.userlog("tpgetlev()={}".format(t.tpgetlev()))
            time.sleep(1)
        t.userlog("tpgetlev()={} before".format(t.tpgetlev()))
        try:
            t.tpcall("PING", {})
        except t.XatmiException as e:
            t.userlog("Got error {}".format(e))
        t.userlog("tpgetlev()={} after".format(t.tpgetlev()))
        return t.tpreturn(t.TPSUCCESS, 0, req)
t.run(Server(), sys.argv)
```

We have a new service called SLEEP that does exactly that. It will sleep for *10* seconds, checking whether the service is still in a transaction. Then, it will act as a client of the PING service, log errors, and finish.

Once you have started the application, we can call the new service in a transaction with a *1*-second timeout:

```
echo -e "SRVCNM\tSLEEP\n" | ud32 -t 1
```

First, unless you won the lottery, it will take more than 1 second before you get an error about Return packet time out. But ULOG will contain even more interesting output:

- The service sleeps for all *10* seconds and it is still in a transaction.

- The call to the PING service fails with a TPETIME error but the PING service was not called.

- The tpreturn function failed because our client exited and removed the reply queue.

We can also verify whether the service succeeds if it is given more than *10* seconds to complete:

```
echo -e "SRVCNM\tSLEEP\n" | ud32 -t 11
```

It will have the expected output in ULOG including the successful call to the PING service in a transaction.

> **Tip**
> Find a way to enforce reasonable time constraints on your *pure* Tuxedo servers. Otherwise, they will be working hard to prepare the result for clients that have already moved on due to timeouts and they will exhaust available server instances.

Now that we have investigated transaction timeouts, let's have a look at the second kind of timeouts.

Blocking timeouts

Most of the time, blocking timeouts are configured globally in the configuration file, since blocking on a system call should be a rare occasion. There is a tpgblktime function call to obtain the current timeout value and a second tpsblktime function to change the blocking timeout. Both functions can be called either for the next potentially blocking call with the TPBLK_NEXT flag or all of them with the TPBLK_ALL flag. The timeout can be configured either in milliseconds with the TPBLK_MILLISECOND flag or in seconds using the TPBLK_SECOND flag. For millisecond timeouts to work, the whole application must be configured with SCANUNIT in milliseconds. And of course, the blocking timeout will be rounded up to the closest multiple of SCANUNIT because it is implemented by periodic scans.

Unfortunately, the tpgblktime function does not return the global blocking timeout, only the value set by the tpsblktime function. So, we have to remember that the global blocking timeout was *15* seconds and the SLEEP service slept for *10* seconds. To experience blocking timeouts, we will have to adjust the blocking timeout by hand:

```
import tuxedo as t
t.tpsblktime(5, t.TPBLK_ALL | t.TPBLK_SECOND)
t.tpcall("SLEEP", {})
```

After 5 to 10 seconds, you will get an exception saying TPETIME and timeout occurred. Since the tpcall function both sends the request and receives a response, sometimes it's useful to break it down and use different timeouts for the sending part and receiving part. This is because sending a request should never block in a well-administered application, but receiving a response will block for a while if the service does something useful:

```
import tuxedo as t
t.tpsblktime(1, t.TPBLK_ALL | t.TPBLK_SECOND)
cd = t.tpacall("SLEEP", {})
t.tpsblktime(5, t.TPBLK_ALL | t.TPBLK_SECOND)
t.tpgetrply(cd)
```

There is one more thing you must know about the blocking timeout – there is a way to skip it. Using a TPNOTIME flag value will make the calls immune to blocking timeouts. Transaction timeouts, however, can still occur. Since the SLEEP service takes 10 seconds to complete, let's use a 5-second blocking timeout and see whether we can avoid receiving a timeout:

```
import tuxedo as t
t.tpsblktime(5, t.TPBLK_ALL | t.TPBLK_SECOND)
t.tpcall("SLEEP", {}, t.TPNOTIME)
```

Although the timeout is set to *5* seconds, the call will succeed after around *10* seconds.

And here is why we postponed the blocking timeouts until we learned about transactions – when you use transactions, your application has two different timeouts at the same time, and that causes a lot of problems in practice. The shortest of the timeouts will occur, but the TPETIME error will not tell which timeout it was. Whenever you change one of them, you must think about the other. And we have not considered database session timeouts, HTTP connection timeouts, and others yet.

> **Tip**
> Pay extra attention to the application's blocking timeouts whenever you increase transaction timeouts. Consider using the TPETIME flag when your application uses transactions.

Summary

In this chapter, we learned how to configure a Tuxedo application to use transactions. We developed a transaction server and created transactions in application code using the ud32 tool and Tuxedo configuration. We learned about two kinds of timeouts supported by Tuxedo and how they interact. And last, but not least, we learned how to escape transactions and timeouts when necessary.

Now you know how to use transactions in Tuxedo and the quirks and glitches of different timeouts. That means that we can move on to the next chapter about persistent message queues, which use transactions and the transaction manager for ACID properties.

Questions

1. What flag value allows you to a call service outside a transaction?

2. What flag value allows you to call a service without blocking a timeout?

3. Which parameter controls how often timeout detection will happen?

4. Can the .TMIB service be called as a part of the transaction ?

Further reading

- About transactions: https://docs.oracle.com/cd/E72452_01/tuxedo/docs1222/ads/adtrn.html

- Configuring transactions: https://docs.oracle.com/cd/E72452_01/tuxedo/docs1222/ads/adtran.html

8
Using Tuxedo Message Queue

As you know by now, Tuxedo uses System V IPC message queues for inter-process communication. Not only are those queues not persistent but their content can also be lost during a system restart. Tuxedo also uses at-most-once delivery semantics. So then, how can one build a reliable system with these tools? The answer to this is to use persistence for some of the messages and application parts.

Tuxedo comes with a queueing component called /Q included. Among other features, it supports queues persisted to the file system. Oracle also offers a paid add-on called **Oracle Tuxedo Message Queue** (**OTMQ**) with even more features. However, in this book, we will cover only /Q because every Tuxedo application has this component and it is sufficient for most tasks. For more advanced features and better integration, we will be using a open source messaging framework in *Chapter 12*, *Modernizing the Tuxedo Application*.

In this chapter, we will cover the following topics:

- Creating and configuring queues
- Using queues
- Forwarding messages

By the end of this chapter, you will have learned how to create persistent queues and the essential queue parameters. You will be able to configure an application and will have learned how to enqueue and dequeue messages in multiple ways. Finally, you will have learned how to use message forwarding to build more reliable applications.

Technical requirements

All code for this chapter is available at `https://github.com/PacktPublishing/Modernizing-Oracle-Tuxedo-Applications-with-Python/tree/main/Chapter08`.

The Code in Action video for the chapter can be found at `https://bit.ly/3qVgdqL`.

Creating and configuring queues

Messages are stored in *a queue* that belongs to *a queue space* that resides in *a device*. Just like the transaction log, a queue device is just a file of a preallocated size and with a special format. First, we start the configuring by specifying a file that will contain the device:

```
export QMCONFIG=`pwd`/qmconfig
```

This environment variable is needed for `qmadmin` to work on the specified device.

Next, we create the actual device using the `qmadmin` tool and specify the device offset and a size of *200* blocks:

```
echo "crdl $QMCONFIG 0 200" | qmadmin
```

After that, we can create a queue space that will contain all of our queues. There are several parameters for queue space creation and you can find more information about them in the Tuxedo documentation. We will create our queue space with the following command:

```
echo "qspc QSPACE 230458 100 3 5 5 100 ERR y 16" | qmadmin
```

This command creates a queue space with the following parameters:

- The name of the queue space is `QSPACE`.
- An IPC key, `230458`, must be a unique value greater than `32768` and `262143`.
- The size of the queue space will be `100` blocks; it must be smaller than the size of the device.

- There will be 3 queues in the queue space.

- 5 concurrent transactions and 5 processes will be supported. These values should be kept in sync and be slightly larger than the number of processes that will access these queues.

- Up to 100 messages can be stored in the queue space.

- The name of the error queue is ERR and we will create it later.

- And we will initialize extents (y) and set the blocking factor to 16.

Then we create the following error queue:

```
echo -e "qopen QSPACE\nqcr ERR time none 0 0 50% 10% ''" |
qmadmin
```

It creates a queue with the following parameters:

- The name of the queue is ERR.

- The queue order will be based on time. There are other options such as fifo for first-in, first-out, and priority for a priority-based queue but time is often good enough.

- Enabling out-of-order enqueueing should be ignored and set to none.

- The next two parameters are the number of retries and the number of seconds between the retries. After retries are exhausted, the message is moved to the error queue. Since we're creating the error queue itself right now, set both values to 0.

- Next, there are settings for queue capacity. Once 50% of the capacity is exceeded, a command will be executed. The command will not be executed again unless the queue size drops below 10% and then exceeds 50% again. We use an empty command here for examples but you should specify one that sends an alert for production systems.

Then we create two more queues:

```
echo -e "qopen QSPACE\nqcr REQ time none 3 5 50% 10% ''" |
qmadmin
echo -e "qopen QSPACE\nqcr RES time none 3 5 50% 10% ''" |
qmadmin
```

The queues are called REQ and RES and both will make 3 retries with the 5-second interval before giving up and moving the message to the ERR queue.

Once we have the queues created, we must configure the application to use them.

First, create the configuration file called ubbconfig with the following content:

```
*RESOURCES
MASTER tuxapp
MODEL SHM
IPCKEY 32769
SCANUNIT 5
BLOCKTIME 3
*MACHINES
"15c365dcb562" LMID=tuxapp
    TUXCONFIG="/home/oracle/code/08/tuxconfig"
    TUXDIR="/home/oracle/tuxhome/tuxedo12.2.2.0.0"
    APPDIR="/home/oracle/code/08"
    TLOGDEVICE="/home/oracle/code/08/tlog"
    TLOGSIZE=100
*GROUPS
GROUP1 LMID=tuxapp GRPNO=1 TMSNAME=TMS_QM TMSCOUNT=2
    OPENINFO="TUXEDO/QM:/home/oracle/code/08/
qmconfig:QSPACE"
*SERVERS
TMQUEUE SRVGRP=GROUP1 SRVID=1
    REPLYQ=Y MAXGEN=2 RESTART=Y GRACE=0
    MIN=1 MAX=1
    CLOPT="-s QSPACE:TMQUEUE -- "
```

For queues to work, a group for the queue server must be configured. The most important parameters are TMSNAME with TMS_QM as the transaction management server. It also makes all queue operations transactional. The second important parameter is OPENINFO, consisting of three parts separated by a colon: TUXEDO/QM, the device file containing the queues, and the queue space name QSPACE.

Then we need to configure the TMQUEUE server that comes with Tuxedo. The most important parameter is CLOPT, which advertises a QSPACE service matching our queue space name. The Tuxedo API functions for enqueueing and dequeuing messages do not work directly with the queues. Instead, the functions call the service provided by TMQUEUE to perform all operations.

The /Q queue API also requires transactions so we must configure the transaction log as we learned in the previous chapter:

```
tmloadcf -y ubbconfig
echo crdl -z `pwd`/tlog -b 200 | tmadmin
echo crlog -m tuxapp | tmadmin
```

Once this is done, we are good to start the application and start using the queues.

Using queues

There is a special parameter for the queue API called the **queue control structure**, which is called TPQCTL in the API. It contains additional parameters and response information for the queue operations. Because of the heritage of the C programming language, each parameter must be paired with a flag value that indicates that the parameter is set. Otherwise, it is ignored. The most important parameters are as follows:

- corrid is a correlation identifier up to 32 characters long. TPQCORRID must be set in flags to indicate the presence of this field.

- deq_time tells when the message should be dequeued from the queue. When TPQTIME_ABS is set in flags, deq_time contains seconds since the UTC epoch. When TPQTIME_REL is set in flags, the value is the number of seconds after enqueueing the message.

- replyqueue and failurequeue are names of queues where a successful or failure response message will be stored. TPQREPLY or TPQFAILUREQ must be set in flags to indicate the presence of these parameters.

The simplest way to store a message in a queue is to call the tpenqueue function. This is shown as follows:

```
import tuxedo as t
qctl = t.tpenqueue(
    "QSPACE", "ERR", t.TPQCTL(), {"TA_STATUS": "Hello /Q!"}
)
```

From the preceding code, we see that the parameters for the `tpenqueue` function are the queue space, queue name, queue control structure, message, and optional flags. The most useful flags are `TPNOTRAN` to store the message outside of the global transaction, `TPNOTIME` to ignore the blocking timeout, and `TPNOBLOCK` to raise an exception immediately if the queue is full. The function returns the queue control structure with some additional information. We can verify that the message is in the queue by using the following `qmadmin` tool to list messages:

```
echo -e "qopen QSPACE\nqset ERR\nql" | qmadmin
```

Now we can dequeue the message using the `tpdequeue` function:

```
qctl, data = t.tpdequeue("QSPACE", "ERR", t.TPQCTL())
```

It takes parameters containing the queue space, queue name, queue control structure, and optional flags. The useful flags are `TPNOTRAN` to store the message outside of the global transaction, `TPNOTIME` to ignore the blocking timeout, and `TPNOBLOCK` to fail immediately if blocking conditions exist. The function returns the queue control structure and the message data dequeued. You can even verify that the data returned is the same as the data enqueued and that `qmadmin` does not show any messages in the queue.

The `tpdequeue` function also supports the following flags in the queue control structure:

- `TPQGETBYMSGID` retrieves the message with the specified message identifier.
- `TPQGETBYCORRID` retrieves the message with the specified correlation identifier.
- `TPQWAIT` will wait for the message with a specified message identifier or correlation identifier to appear in the queue.
- `TPQPEEK` reads the message but does not remove it from the queue.

A more realistic use of persistent queues requires filling the queue control structure. Often, you have to assign a unique identifier to the request message so you can match it with the response message. A correlation identifier can be used for that. Here, we use `hello` as the correlation identifier and we must reflect that in `flags` by setting `TPQCORRID`. You can also set the `TPQMSGID` flag to retrieve the unique message identifier assigned by Tuxedo, the same one you saw in the output of `qmadmin`. Here is the code to do that:

```
qctl = t.tpenqueue(
    "QSPACE",
    "ERR",
    t.TPQCTL(corrid="hello", flags=t.TPQCORRID + t.TPQMSGID),
```

```
    {"TA_STATUS": "Hello!"},
)
print(qctl.msgid)
```

You can also use the correlation identifier to retrieve a specific message instead of one according to queue ordering configuration by using the queue control structure for the `tpdequeue` function:

```
qctl, data = t.tpdequeue(
    "QSPACE",
    "ERR",
    t.TPQCTL(corrid="hello", flags=t.TPQGETBYCORRID),
)
```

Another good use of queues is to schedule a delayed execution. We can do that by enqueueing a message that will appear in the queue after a specified time. We use 60 seconds in the following example:

```
qctl = t.tpenqueue(
    "QSPACE",
    "ERR",
    t.TPQCTL(deq_time=60, flags=t.TPQTIME_REL),
    {"TA_STATUS": "Delayed hello!"},
)
```

Assume that you need to dequeue a message like the following:

```
qctl, data = t.tpdequeue("QSPACE", "ERR", t.TPQCTL())
```

The dequeing will fail with an exception. You have to wait for 60 seconds and repeat the dequeue operation to receive the message successfully.

So far, we have used queues programmatically, but Tuxedo comes with more tools for working with queues and we will look at those next.

Forwarding messages

Tuxedo comes with a TMQFORWARD server that dequeues a message from the queue and sends it to a service with the same name as the queue. It also uses the number of retries and the interval between retries we configured for the queues to perform retries if the service fails.

We configured the queues to do 3 retries with a 5-second interval. That enables you to create a store and forward solution that works with unreliable destinations and can tolerate temporary network failures or downtime during system upgrades.

Several highly available and reliable systems are built by having persistent request and response queues on the system boundary. Each request is persistent and then processed by the system. If the system fails, the request is retried. Once it succeeds, the request is removed from the queue and the response is persisted. We will use Tuxedo to recreate this solution now.

We start by updating our ubbconfig file with the highlighted lines:

```
*RESOURCES
MASTER tuxapp
MODEL SHM
IPCKEY 32769
SCANUNIT 5
BLOCKTIME 3
*MACHINES
"15c365dcb562" LMID=tuxapp
    TUXCONFIG="/home/oracle/code/08/tuxconfig"
    TUXDIR="/home/oracle/tuxhome/tuxedo12.2.2.0.0"
    APPDIR="/home/oracle/code/08"
    TLOGDEVICE="/home/oracle/code/08/tlog"
    TLOGSIZE=100
*GROUPS
GROUP1 LMID=tuxapp GRPNO=1 TMSNAME=TMS_QM TMSCOUNT=2
    OPENINFO="TUXEDO/QM:/home/oracle/code/08/qmconfig:QSPACE"
GROUP2 LMID=tuxapp GRPNO=2 TMSNAME=TMS TMSCOUNT=2
*SERVERS
TMQUEUE SRVGRP=GROUP1 SRVID=1
    REPLYQ=Y MAXGEN=2 RESTART=Y GRACE=0
    MIN=1 MAX=1
```

```
    CLOPT="-s QSPACE:TMQUEUE -- "
"proc.py" SRVGRP=GROUP2 SRVID=3
    REPLYQ=Y MAXGEN=2 RESTART=Y GRACE=0
    MIN=1 MAX=1
TMQFORWARD SRVGRP=GROUP1 SRVID=3
    REPLYQ=N MAXGEN=2 RESTART=Y GRACE=0
    MIN=1 MAX=1
    CLOPT="-A -- -q REQ -i 1"
```

The TMQFORWARD server must be configured in a group with TMS_QM as the transaction management server. The CLOPT parameter contains the following command-line parameters:

- -q instructs the server to read messages from the queue named REQ to forward them to the service with the same name.

- -i instructs to server look for new messages each second. You can use the value 0 to reduce the latency but this increases the load on the system by scanning the queue all the time.

The proc.py file is our application server with the following content:

```
#!/usr/bin/env python3
import sys
import tuxedo as t
class Server:
    def tpsvrinit(self, argv):
        t.tpadvertise("REQ")
        return 0
    def REQ(self, req):
        req["TA_STATUS"] = "done"
        return t.tpreturn(t.TPSUCCESS, 0, req)
t.run(Server(), sys.argv)
```

It has a single REQ service that updates the response with TA_STATUS equal to done.

Now you can start the reconfigured application. We will enqueue the request into the REQ queue since the TPQFORWARD server is processing it. We assign a unique message correlation identifier, 8, to retrieve it later. And we also specify the reply queue RES where we will expect the responses. Refer to the following code:

```
import tuxedo as t
qctl = t.tpenqueue(
    "QSPACE",
    "REQ",
    t.TPQCTL(
        corrid="8",
        replyqueue="RES",
        flags=t.TPQCORRID + t.TPQREPLYQ,
    ),
    {"TA_STATUS": "do"},
)
```

Then we can start waiting for the response in the reply queue by using the same correlation identifier. Notice that we use the TPQWAIT flag to wait for the response instead of failing when the message is not found:

```
import tuxedo as t
qctl, data = t.tpdequeue(
    "QSPACE",
    "RES",
    t.TPQCTL(corrid="8", flags=t.TPQGETBYCORRID + t.TPQWAIT),
)
print(data)
```

The response will be available in a short moment. If the proc.py server is stopped or crashes, TMQFORWARD will take care of it by retrying service calls.

Summary

In this chapter, we learned how to use the Tuxedo /Q queue component. We learned how to create a queue device and the queue space needed for creating the actual queues. We learned different ways of enqueueing and dequeuing messages, both with the application code and using the TPQFORWARD server.

After this chapter, you should be able to do basic queue administration. You should also know how to use queues to create a more reliable application and decouple parts of your application. The main limitation of /Q queues is that they are filesystem-based and special care must be taken to ensure the queue device is not lost during a system crash or major upgrade. The other limitation is that these queues are tied to Tuxedo and can't be used from non-Tuxedo applications. So keep /Q for interacting between parts of the Tuxedo application and we will learn how to use more open queue systems in *Chapter 12, Modernizing the Tuxedo Application*. But the next chapter will look at using a database for data persistence and integration with other systems.

Questions

After reading this chapter, you should be able to answer the following questions:

1. Which function is used to enqueue a message?
2. Which function is used to dequeue a message?
3. Which message identifier can be assigned by the application?
4. Which flag value causes the tpdequeue to wait for the specified message identifier?

Further reading

- Documentation of qmadmin: (https://docs.oracle.com/cd/E72452_01/tuxedo/docs1222/rfcm/rfcmd.html).
- Administration of /Q queues: (https://docs.oracle.com/cd/E72452_01/tuxedo/docs1222/qgd/qadm.html).

9
Working with Oracle Database

These days it is hard to find an application that does not use a database for persistence. It might be a relational database, graph database, NoSQL database, or even an SQLite database on your mobile phone. Tuxedo applications are no different. While many databases can be used with Tuxedo, its best friend is the Oracle Database. Tuxedo is optimized for working with the Oracle Database. Even the transaction log we learned about in *Chapter 7*, *Distributed Transactions*, can be stored inside the Oracle Database instead of a file.

You will need an Oracle Database for this chapter. You may use the one you have already or use a Docker image available from the Docker Hub at `https://hub.docker.com/_/oracle-database-enterprise-edition`. Getting access to the Oracle Database is your homework before reading the chapter, as also mentioned in the sections ahead.

In this chapter, we will cover the following topics:

- Preparing the database
- Using local transactions with Oracle Database
- Using global transactions with Oracle Database

By the end of this chapter, you will have learned how to use both local and global transactions with Tuxedo and the Oracle Database. You will learn how to build a transaction management server needed for global transactions and how to configure it. You will also learn a few tips and tricks along the way.

Technical requirements

The Oracle Database is required. You can either use an existing one or download one from Docker Hub and set it up according to the instructions here: `https://hub.docker.com/_/oracle-database-enterprise-edition`.

All code for this chapter is available at `https://github.com/PacktPublishing/Modernizing-Oracle-Tuxedo-Applications-with-Python/tree/main/Chapter09`.

The Code in Action video for the chapter can be found at `https://bit.ly/30RhNzj`.

Preparing the database

At this point, you should have an Oracle Database running on a server or inside a Docker container. We can proceed by making sure the Tuxedo application can access the database. To be able to connect to a remote Oracle Database, we have to install the following Oracle Database client libraries:

```
sudo yum -y install oracle-instantclient19.9-basic
sudo yum -y install oracle-instantclient19.9-devel
sudo yum -y install oracle-instantclient19.9-sqlplus
```

Once the libraries have been successfully installed, we can install the `cx_Oracle` library for accessing the database from Python as follows:

```
sudo pip3 install cx_oracle
```

We have everything ready for writing the code now but we need to do some preparations inside the database as well. If you already have a database user you want to use for the application, skip all commands until step 3. Otherwise, let's create a new database user as follows:

1. We need a regular database user for our application. To create one, you need database administrator credentials along with the hostname and port number. For the Oracle Database Docker container, you can connect to it using `sqlplus` as follows:

    ```
    sqlplus sys/Oradoc_db1@//host.docker.internal:32769/
    ORCLPDB1.localdomain as sysdba
    ```

2. Once you are connected, the following commands will create a user named `tuxedo` for our programming needs:

    ```
    CREATE USER tuxedo IDENTIFIED BY tuxpass;
    GRANT CREATE SESSION TO tuxedo;
    GRANT CREATE TABLE TO tuxedo;
    ALTER USER tuxedo QUOTA UNLIMITED ON users;
    ```

3. When the new user is created, connect to the database with the user credentials. For the Docker container, that will be the following command:

    ```
    sqlplus tuxedo/tuxpass@//host.docker.internal:32769/
    ORCLPDB1.localdomain
    ```

4. Then create a simple table that we will use in our experiments:

    ```
    CREATE TABLE messages (
        message_id VARCHAR2(36),
        content VARCHAR2(256),
        PRIMARY KEY(message_id)
    );
    ```

Now the database is ready and we can start using it from a Tuxedo application.

Using local transactions with Oracle Database

We will start by using local (regular) transactions in a Tuxedo application. To keep it simple, the application will count the number of records in the messages table. This task is not interesting by itself and there are plenty of great resources about using the cx_Oracle library. While there is nothing specific about the local cx_Oracle transactions in Tuxedo, it will serve us as a great contrast when compared to global transactions.

Every Tuxedo application needs a configuration file and we will start with the following ubbconfig file:

```
*RESOURCES
MASTER tuxapp
MODEL SHM
IPCKEY 32769
SCANUNIT 5
BLOCKTIME 3
*MACHINES
"15c365dcb562" LMID=tuxapp
    TUXCONFIG="/home/oracle/code/09/tuxconfig"
    TUXDIR="/home/oracle/tuxhome/tuxedo12.2.2.0.0"
    APPDIR="/home/oracle/code/09"
*GROUPS
GROUP1 LMID=tuxapp GRPNO=1 TMSNAME=TMS TMSCOUNT=2
*SERVERS
"db.py" SRVGRP=GROUP1 SRVID=1
    REPLYQ=Y MAXGEN=2 RESTART=Y GRACE=0
    MIN=1 MAX=1
```

The db.py Tuxedo server will have a single service called COUNT that returns the number of records in the messages table. During server startup, we establish a connection to the Oracle Database in the tpsvrinit function. The connection uses the same parameters as the sqlplus commands before and you may need to adjust them. The COUNT service then uses the connection to execute a SQL query and returns the result in the TA_STATUS field of the response as shown here:

```
#!/usr/bin/env python3
import sys
import tuxedo as t
import cx_Oracle
```

```
class Server:
    def tpsvrinit(self, argv):
        t.tpadvertise("COUNT")
        self.db = cx_Oracle.connect(
            "tuxedo",
            "tuxpass",
            "host.docker.internal:32769/ORCLPDB1.localdomain",
        )
        return 0
    def COUNT(self, req):
        with self.db.cursor() as dbc:
            dbc.execute("SELECT COUNT(1) FROM messages")
            req["TA_STATUS"] = "Count={}".format(dbc.fetchone()
[0])
        return t.tpreturn(t.TPSUCCESS, 0, req)
t.run(Server(), sys.argv)
```

If you have written any `cx_Oracle` code before, the code of `db.py` will not surprise you. We can verify that this service works by calling it as follows:

```
import tuxedo as t
_, _, res = t.tpcall("COUNT", {})
```

The response should contain the `TA_STATUS` field with the value `Count=0` as we don't have any records in the table yet.

Keep in mind, this is example code. To make it robust and error-prone for production, you should consider and address at least the following points:

- Where and how you store database credentials.
- If the database is temporarily inaccessible, `tpsvrinit` will fail, the server will not start, and will not be restarted by Tuxedo. Consider connecting lazily inside the service.
- If the database connection is lost during service execution, you must either reconnect or at least return `TPEXIT` and Tuxedo will restart the server.

In this section, we saw that using local transactions is easy and does not rely on Tuxedo features. Now let's look at how to use the transaction monitor capabilities of Tuxedo.

Using global transactions with Oracle Database

To participate in global transactions, we need a transaction management server that we can specify in the `ubbconfig` GROUP configuration. So far, we have used the TMS *dummy* server and TMS_QM for Tuxedo queues. These servers come bundled with Tuxedo. For the Oracle Database, we have to build one ourselves. Tuxedo provides the `buildtms` tool for this purpose:

```
export ORACLE_HOME=/usr/lib/oracle/19.9/client64
buildtms -v -r Oracle_XA -o TMS_ORA
```

This command will create a new server named TMS_ORA. The `-v` command-line parameter will show you what this command does under the hood by using configuration from the `$TUXDIR/udataobj/RM` file.

The transaction management server we just created needs additional privileges for the database user. We have to connect as an administration once again, as follows:

```
sqlplus sys/Oradoc_db1@//host.docker.internal:32769/ORCLPDB1.
 localdomain as sysdba
```

Then we grant the SELECT privilege on the `dba_pending_transactions` table to our user as follows:

```
GRANT SELECT ON dba_pending_transactions TO tuxedo;
```

This table contains the prepared transactions that were not committed or aborted. Transaction management server uses this table during the startup to perform transaction recovery. Once privileges are granted, we can continue with adjusting the configuration for the application.

> **Tip**
> If TMS_ORA fails to boot, look for errors in files with names similar to `xa_NULL02172021.trc`. Mentions of `xaorecover` and `ORA-00942: table or view does not exist` indicate missing privileges.

The updated `ubbconfig` will have the following content:

```
*RESOURCES
MASTER tuxapp
MODEL SHM
IPCKEY 32769
SCANUNIT 5
BLOCKTIME 3
*MACHINES
"15c365dcb562" LMID=tuxapp
    TUXCONFIG="/home/oracle/code/09/tuxconfig"
    TUXDIR="/home/oracle/tuxhome/tuxedo12.2.2.0.0"
    APPDIR="/home/oracle/code/09"
    TLOGDEVICE="/home/oracle/code/09/tlog"
    TLOGSIZE=100
*GROUPS
GROUP1 LMID=tuxapp GRPNO=1 TMSNAME=TMS TMSCOUNT=2
GROUP2 LMID=tuxapp GRPNO=2 TMSNAME=TMS_ORA TMSCOUNT=2
    OPENINFO="Oracle_XA:Oracle_XA+Acc=P/tuxedo/
tuxpass+SqlNet=//host.docker.internal:32769/ORCLPDB1.
localdomain+SesTm=30+Threads=true+NoLocal=true"
*SERVERS
"db.py" SRVGRP=GROUP1 SRVID=1
    REPLYQ=Y MAXGEN=2 RESTART=Y GRACE=0
    MIN=1 MAX=1
"dbxa.py" SRVGRP=GROUP2 SRVID=2
    REPLYQ=Y MAXGEN=2 RESTART=Y GRACE=0
    MIN=1 MAX=1
```

The application will use global transactions and we must configure a transaction log for it. Then we have a new group with `TMS_ORA` as a transaction manager and several configuration parameters inside `OPENINFO`, such as the following:

- `Acc=P/tuxedo/tuxpass` specifies the username and password for the database connection. If you do not want to have a password in the configuration file, you can put `*****` (five asterisks) instead and the `tmboot` command will ask for the password and store it in encrypted form.

- `SqlNet=//host.docker.internal:32769/ORCLPDB1.localdomain` specifies the Oracle Net database link.

- `SesTm=30` specifies the maximum number of seconds the transaction can be inactive before it is aborted.

- `Threads=true` is used to support multi-threaded applications that we will develop later.

- `NoLocal=true` forbids local transactions. If the code ignores transaction exceptions and continues to use the database cursor, a local transaction might be started and it is a common source of errors. So it is better to disable local transactions.

> **Tip**
> Now you have to adjust three timeouts in sync: blocking timeout, transaction timeout, and Oracle database session timeout. Otherwise, the smallest one will occur unexpectedly.

As you should remember from *Chapter 7, Distributed Transactions*, you have to perform the following extra steps to create a transaction log:

```
tmloadcf -y ubbconfig
echo crdl -z `pwd`/tlog -b 200 | tmadmin
echo crlog -m tuxapp | tmadmin
```

Then it is time to develop a Tuxedo server that participates in the global transaction. We will call it `dbxa.py` and it will contain the following code:

```python
#!/usr/bin/env python3
import sys
import tuxedo as t
import cx_Oracle
import uuid
class Server:
    def tpsvrinit(self, argv):
        t.tpadvertise("STORE")
        t.tpadvertise("COUNTXA")
        self.db = cx_Oracle.connect(handle=t.xaoSvcCtx())
        return 0
    def STORE(self, req):
```

```
            with self.db.cursor() as dbc:
                dbc.execute(
                    "INSERT INTO messages VALUES (:1, :2)",
                    [str(uuid.uuid4()), req["TA_SOURCE"][0]],
                )
            return t.tpreturn(t.TPSUCCESS, 0, req)
        def COUNTXA(self, req):
            with self.db.cursor() as dbc:
                dbc.execute("SELECT COUNT(1) FROM messages")
                req["TA_STATUS"] = "Count={}".format(dbc.fetchone()
[0])
            return t.tpreturn(t.TPSUCCESS, 0, req)
t.run(Server(), sys.argv, "Oracle_XA")
```

This code contains two services. The COUNTXA service is the exact copy of the COUNT service and no changes are needed for it to work in a global transaction. The STORE service adds a new record to the messages table. Notice that the service lacks a call commit or rollback functions: that will happen during the commit or the aborting of the global transaction. In the case of local transactions, a commit or rollback would be mandatory.

The main difference here compared to the local transaction example is that the connection is created without a database address and credentials. Instead, a handle to xaoSvcCtx is given. Tuxedo will do its magic to establish a database connection with the configuration provided in the group's OPENINFO parameter under the hood. The last but not least change is to provide the resource manager name Oracle_XA as the third parameter to the run function. Otherwise, the server will fail during startup.

We can start the Tuxedo application now and examine our services:

```
import tuxedo as t
t.tpbegin(30)
t.tpcall("STORE", {"TA_SOURCE": "Hello"})
```

We started a global transaction with a 30-second timeout and stored a new message in the table. Let's call the COUNT service we developed before:

```
t.tpcall("COUNT", {}).data
```

Although the service participates in the global transaction, it will return `Count=0`. While it shares the global Tuxedo transaction, it does not share the underlying Oracle Database transaction. Now contrast that with the `COUNTXA` service as follows:

```
t.tpcall("COUNTXA", {}).data
```

It will return `Count=1` because it participates in the database transaction. You might be tempted to assume this is because `COUNTXA` and `STORE` reside in the same Tuxedo server, but no. You could do another exercise and split `dbxa.py` into two servers each containing one of the services and the result would still be the same. At this point, we can complete the transaction as follows:

```
t.tpcommit()
```

Assuming it took you less than 30 seconds from executing `tpbegin` till executing `tpcommit`, it will succeed. If it took more time, there will be a transaction timeout and `tpcommit` or any preceding command will fail. You have to repeat the code and be faster the second time. Once the transaction succeeds, we can call the `COUNT` service once again:

```
t.tpcall("COUNT", {}).data
```

It will return `Count=1` because the changes have been persisted and made available to everyone.

> **Tip**
> Unlike local transactions where we have to handle connectivity errors and restore the connection, it is done implicitly behind the scenes for servers using XA connection and global transactions.

We have completed a global transaction with multiple services participating in it. There was one topic I promised to write about while explaining the `OPENINFO` parameters: threads.

Multi-threading and global transactions

The first rule of multi-threading is: you do not use multi-threading. Tuxedo works best with single-threaded servers. Single-threaded servers cannot corrupt other servers' memory. Threads can corrupt other threads' memory. Not to mention that multiple threads do not work great in Python due to the **Global Interpreter Lock**. But there are exceptions to every rule and sometimes you need to have multiple threads.

You will need a connection object for each thread and the best way to do that is by creating it in the `tpsvrthrinit` function that is called once for each thread of the Tuxedo server and to store it in the thread-local storage of `threading.local`. Refer to the following code:

```python
#!/usr/bin/env python3
import sys
import tuxedo as t
import cx_Oracle
import uuid
import threading
class Server:
    def tpsvrinit(self, argv):
        t.tpadvertise("STORE")
        t.tpadvertise("COUNTXA")
        self.local = threading.local()
        return 0
    def tpsvrthrinit(self, argv):
        self.local.db = cx_Oracle.connect(
            handle=t.xaoSvcCtx(), threaded=True
        )
        return 0
    def STORE(self, req):
        with self.local.db.cursor() as dbc:
            dbc.execute(
                "INSERT INTO messages VALUES (:1, :2)",
                [str(uuid.uuid4()), req["TA_SOURCE"][0]],
            )
        return t.tpreturn(t.TPSUCCESS, 0, req)
    def COUNTXA(self, req):
        with self.local.db.cursor() as dbc:
            dbc.execute("SELECT COUNT(1) FROM messages")
            req["TA_STATUS"] = "Count={}".format(dbc.fetchone()
[0])
        return t.tpreturn(t.TPSUCCESS, 0, req)
t.run(Server(), sys.argv, "Oracle_XA")
```

The only change to the configuration is to tell Tuxedo how many threads should be started. This is shown in the following code:

```
"dbxa.py" SRVGRP=GROUP2 SRVID=2
    REPLYQ=Y MAXGEN=2 RESTART=Y GRACE=0
    MIN=1 MAX=1
    MINDISPATCHTHREADS=4 MAXDISPATCHTHREADS=4
```

And now you have a multi-threaded Tuxedo server doing global transactions in the Oracle Database. There are many things to learn about using databases but we have covered most if not all that you have to know to use Oracle Database from a Tuxedo application.

Summary

In this chapter, we used the Oracle Database both from local and global transactions. You also experienced that mixing services using both local and global transactions does not work as you might imagine at first. Global transactions worked without any effort in application code and you might wonder why you would ever need to use local transactions. That feeling is shared by many users of Tuxedo. The most common reason for not using global transactions is squeezing out the last bits of performance. The performance overhead of transactions is small but they are not free. But even in those cases, it pays to have two implementations of the service as we did with the COUNT and COUNTXA services, one for each transaction type. After all, the code is the same and the only difference is in the server initialization.

Now you know how to develop servers for global and local transactions and how to configure the Tuxedo application in each case. You also know how to access the database from a multi-threaded server in case you need it badly. You've had hands-on experience of mixing local and global transactions, which leaves many scratching their heads and wondering where their data went. Hopefully, you will not experience that anymore.

Up until now, we looked at the Tuxedo application in isolation and used it from the same machine. But in real-life, a Tuxedo application has to be integrated with the rest of the infrastructure. In the next chapter, we will learn how to access the Tuxedo application over the network from your web server, desktop application, or other pieces of software.

Questions

1. What is the command for building a transaction management server?

2. On which table is the SELECT privilege required for global transactions to work?

3. How can you avoid storing the password in the ubbconfig file and make tmloadcf ask for it?

Further reading

- Transaction log in Oracle Database: https://docs.oracle.com/cd/ E53645_01/tuxedo/docs12cr2/ads/adtrn.html

- Using Oracle XA with transaction monitors: https://docs.oracle.com/cd/ B12037_01/appdev.101/b10795/adfns_xa.htm

- The documentation of cx_Oracle: https://cx-oracle.readthedocs.io/ en/latest/

Section 3:
Integrations

The Tuxedo application is not an isolated island; it is a part of a larger information system and has to interact with other non-Tuxedo applications. We will learn about incoming connections and outgoing connections, and then mix them all by integrating with NATS applications.

This section has the following chapters:

- *Chapter 10, Accessing the Tuxedo Application*
- *Chapter 11, Consuming External Services in Tuxedo*
- *Chapter 12, Modernizing the Tuxedo Application*

10
Accessing the Tuxedo Application

Every Tuxedo client we have developed so far was running on the same physical machine. But that is a rare occasion in practice. Most applications have a frontend either running on a desktop computer or web-based interfaces running in browsers. The backend Tuxedo application is running on a big server behind firewalls. The frontend and backend must communicate with each other. In this chapter, we will learn what Tuxedo offers to solve this task of accessing the application over the network. We will also see what we can create with a small effort by using Python's libraries.

In this chapter, we will cover the following topics:

- Using the Tuxedo Workstation client
- Exposing Tuxedo services as web services

By the end of this chapter, you will have learned how to access the backend Tuxedo application from a different frontend machine over the network. You will learn how to use the Tuxedo Workstation component to access the powerful XATMI API. With that under your belt, you will learn how to create web services that call the Tuxedo application.

Technical requirements

All code for this chapter is available at `https://github.com/PacktPublishing/Modernizing-Oracle-Tuxedo-Applications-with-Python/tree/main/Chapter10`.

The Code in Action video for the chapter can be found at `https://bit.ly/38R4TFP`.

Using the Tuxedo Workstation client

Throughout the chapters so far, we have used the Tuxedo XATMI API for developing clients and accessing the application running on the same machine. But what if we need to access the Tuxedo application from a different machine over the network? The answer to that is the Workstation component that Tuxedo offers out of the box. The main benefit is that you can develop remote clients using the same API as local (native) clients and you have access to the same features including calling any service, starting and finishing transactions.

Let's see how this works with a simple application and an `ubbconfig` file with the following content:

```
*RESOURCES
MASTER tuxapp
MODEL SHM
IPCKEY 32769
*MACHINES
"15c365dcb562" LMID=tuxapp
    TUXCONFIG="/home/oracle/code/10/tuxconfig"
    TUXDIR="/home/oracle/tuxhome/tuxedo12.2.2.0.0"
    APPDIR="/home/oracle/code/10"
*GROUPS
GROUP1 LMID=tuxapp GRPNO=1 TMSNAME=TMS TMSCOUNT=2
*SERVERS
"ping.py" SRVGRP=GROUP1 SRVID=1
    REPLYQ=Y MAXGEN=2 RESTART=Y GRACE=0
    MIN=1 MAX=1
```

The configuration file contains a single `ping.py` server we've used several times already with the following content:

```python
#!/usr/bin/env python3
import sys
import tuxedo as t
class Server:
    def tpsvrinit(self, argv):
        t.tpadvertise("PING")
        return 0
    def PING(self, req):
        req["TA_STATUS"] = "v1"
        return t.tpreturn(t.TPSUCCESS, 0, req)
t.run(Server(), sys.argv)
```

Once we have created the files, we can start the application and access it from a client as follows:

```python
import tuxedo as t
t.tpcall("PING", {}).data
```

Have you ever wondered how the client program knows how to connect to the application? Even though the client is running on the same machine, the client must know how to discover which services are available and what the services' request queues are. The answer to this question is the TUXCONFIG environment variable. Just like the Tuxedo application and its servers, clients use the tuxconfig file pointed to by the TUXCONFIG environment variable to discover the application's configuration. We can remove the value of TUXCONFIG to see that the client does not work. Just start the Python interpreter with the following command:

```
TUXCONFIG="" python3
```

If you execute the following code again, you will get an error during the service call:

```python
import tuxedo as t
t.tpcall("PING", {}).data
```

Accessing the Tuxedo application over the network from a different machine does not work out of the box. We will have to change the application configuration. Let's see how this is done.

Configuring Workstation Listener

To access the Tuxedo application over the network, we must configure a Tuxedo server called **Workstation Listener**. We have to update our ubbconfig with the following highlighted configuration:

```
*RESOURCES
MASTER tuxapp
MODEL SHM
IPCKEY 32769
MAXACCESSERS 50
*MACHINES
"15c365dcb562" LMID=tuxapp
     TUXCONFIG="/home/oracle/code/10/tuxconfig"
     TUXDIR="/home/oracle/tuxhome/tuxedo12.2.2.0.0"
     APPDIR="/home/oracle/code/10"
     MAXWSCLIENTS=10
*GROUPS
GROUP1 LMID=tuxapp GRPNO=1 TMSNAME=TMS TMSCOUNT=2
*SERVERS
"ping.py" SRVGRP=GROUP1 SRVID=1
     REPLYQ=Y MAXGEN=2 RESTART=Y GRACE=0
     MIN=1 MAX=1
WSL SRVGRP=GROUP1 SRVID=10
     REPLYQ=Y MAXGEN=2 RESTART=Y GRACE=0
     MIN=1 MAX=1
     CLOPT="-A -- -n //0.0.0.0:8000 -m 3 -M 3"
```

One of the changes here is a new MAXWSCLIENTS parameter for the machine. It specifies the number of workstation clients that can be connected to the application. MAXWSCLIENTS uses slots reserved by the MAXACCESSERS parameter and you must ensure that MAXACCESSERS is larger than the sum of all clients, servers, and MAXWSCLIENTS.

Next, we must add the WSL server that will listen on a TCP port and offer access to the application over the network. There are many configuration parameters described in the documentation, but we will look at the most important ones:

- -n specifies the network address to listen for incoming connections. The value //0.0.0.0:8000 will make WSL listen on port 8000 of all network interfaces. This is also the only required parameter.

- -m and -M specifies the minimum and the maximum number of handlers. The WSL server itself does not handle the clients and starts another process called WSH to serve requests from the clients.

- -p and -P specifies the minimum and maximum port number used by WSH. The connection will be only initially made to the port given in the -n parameter. After the initial connection to the WSL server, the clients will connect to the WSH process using a port number within an interval given by the -p and -P parameters.

The WSL server has multiple parameters to enable the encryption of connections. But I strongly recommend using external tools and encrypted overlay networks for that. You will have more control over the encryption algorithms used and a faster response to vulnerabilities without patching and upgrading the application.

Once you have started the application, let's see what happens:

1. Here are the started processes:

```
ps aux | grep WS
```

You can see here the WSL listener process and three WSH handler processes that were started behind the scenes.

2. For the Workstation client to connect to the Tuxedo application, we must specify the network address of the WSL listener in the WSNADDR environment variable:

```
export WSNADDR=//localhost:8000
```

We will connect to the application on the same server through the network. But you can do it from a different machine as well if you install Tuxedo on it. The Tuxedo installation has an option to install just the client libraries that will be enough to connect to WSL if you care about the disk usage. But we can start the Python interpreter with TUXCONFIG on the same machine to ensure the local connection is not used:

```
TUXCONFIG="" python3
```

The only thing we have to change in our client code is to use the tuxedowsc library instead of tuxedo. Unfortunately, Tuxedo makes it impossible to support both the local and the remote connection in the same library. But at least the API is the same:

```
import tuxedowsc as t
t.tpcall("PING", {}).data
```

And there you go – we just called the Tuxedo application over the TCP network. You can also check the active connections in the same Python interpreter using the following code:

```
import os
os.system("ss -p")
```

This will show that our Python interpreter is connected to the WSH process, not the WSL process on a port within the configured range.

> **Tip**
>
> The need to expose the WSL port and any port that may be used by the WSH processes causes pain for network administrators. It is better to configure a larger number of handler processes and ports in advance.

We have achieved our goal of connecting to the application over the network. While it is convenient for developers familiar with Tuxedo, it will not be familiar for most developers used to XML and JSON web services. Let's see if there is a more modern way of accessing the Tuxedo application.

Exposing Tuxedo services as web services

The Workstation client API is limited to programming languages that provide bindings for Tuxedo libraries. But not everyone wants to learn a new paradigm and API just to interact with a Tuxedo application, although this book attempts to teach it. Most developers are familiar with the HTTP protocol and XML or JSON data formats used by web services.

There are several solutions to expose Tuxedo services as web services:

- Oracle Tuxedo offers a paid add-on called **Oracle SALT** that enables you to expose Tuxedo services as web services with a SOAP, XML, JSON, and HTML payload and call web services from the Tuxedo application. This add-on is attractive if your tool is the C programming language without standard libraries for high-level network programming. With Python, you have plenty of libraries and Stack Overflow at your service.

- Another solution is to use Flask, FastAPI, or any other Python framework to develop the API, and provide OpenAPI specification and API documentation. The API implementation can call the Tuxedo application using either the native or the Workstation client.

- My favorite solution is to run an API framework in a Tuxedo server. It requires some clever code in older versions of Tuxedo, but current versions of Tuxedo have a new API that simplifies the task. And the `tuxedo` library takes care of tricky details. The main benefit of using Tuxedo servers is that web services can be stopped and started along with the Tuxedo application. A Tuxedo server can provide services that the application can use to push updates over the WebSocket API and more. And the final reason is that Oracle SALT uses this approach as well.

The main limitation of web services is the lack of transactions, which, on the other hand, is not a problem in practice. The service exposed may be an aggregate service that starts a transaction, calls multiple services, and finishes the transaction. Having multiple transactions is something Tuxedo developers are used to, not the rest of the world.

We will create a Tuxedo server with a web server inside of it. For that, we have to update the `ubbconfig` file:

```
*RESOURCES
MASTER tuxapp
MODEL SHM
IPCKEY 32769
*MACHINES
"15c365dcb562" LMID=tuxapp
    TUXCONFIG="/home/oracle/code/10/tuxconfig"
    TUXDIR="/home/oracle/tuxhome/tuxedo12.2.2.0.0"
    APPDIR="/home/oracle/code/10"
*GROUPS
GROUP1 LMID=tuxapp GRPNO=1 TMSNAME=TMS TMSCOUNT=2
*SERVERS
"ping.py" SRVGRP=GROUP1 SRVID=1
    REPLYQ=Y MAXGEN=2 RESTART=Y GRACE=0
    MIN=1 MAX=1
"api.py" SRVGRP=GROUP1 SRVID=20
    REPLYQ=Y MAXGEN=2 RESTART=Y GRACE=0
    MIN=1 MAX=1
```

We no longer have the `WSL` server, but we have a new server, `api.py`, that will provide remote access to the application. The `api.py` server has the following content:

```
#!/usr/bin/env python3
import sys
```

```python
import http.server
import socketserver
import json
import threading
import tuxedo as t
class Handler(http.server.BaseHTTPRequestHandler):
    def do_POST(self):
        content = self.rfile.read(
            int(self.headers.get("Content-Length", "0"))
        )
        req = json.loads(content)
        _, _, res = t.tpcall(self.path[1:], req)

        self.send_response(200)
        self.send_header("Content-type", "application/json")
        self.end_headers()
        self.wfile.write(json.dumps(res).encode("utf-8"))
def serve():
    socketserver.ThreadingTCPServer.allow_reuse_address = True
    with socketserver.ThreadingTCPServer(
        ("", 8000), Handler
    ) as httpd:
        httpd.serve_forever()
class Server:
    def tpsvrinit(self, args):
        threading.Thread(target=serve, daemon=True).start()
        return 0
if __name__ == "__main__":
    t.run(Server(), sys.argv)
```

From the preceding code, we see that during server startup, the tpsvrinit function is called, which starts a new background thread running a server. The serve function starts a web server using the socketserver library just because it comes bundled with Python. It might use any other library of your choice. The request handler reads the content received in the POST request and parses JSON to a Python dictionary that can be used for Tuxedo service requests. The service name is taken from the HTTP request path. The service is invoked and the response is returned as JSON again.

The code of `api.py` looks trivial at first, but the magic happens during `tpcall` when the `tuxedo` library joins the Tuxedo application from a thread created by Python code. It is achieved by using the `tpappthrinit` function, which was added in Oracle Tuxedo 10.3 under the hood. For older versions of Tuxedo, it was not possible to create a thread that calls Tuxedo services. You would have to call a dummy service implemented in the same server to steal a thread with a connection to Tuxedo.

However, now we can start the application and test it using cURL:

```
curl --data '{}' localhost:8000/PING
```

You will get a response from the `PING` service with the `TA_STATUS` field and the value `v1`. We can even invoke the `.TMIB` service and query information about the Tuxedo application as follows:

```
curl --data '{"TA_CLASS":"T_DOMAIN","TA_OPERATION":"GET"}'
localhost:8000/.TMIB
```

We have created our own `ud32` tool for web services! Of course, this web service is too simple and insecure to be exposed to the internet. But still, we implemented some functionality of the SALT add-on in very little time by using Python. Just imagine what could be done given more time!

Summary

In this chapter, we learned how to access the Tuxedo application remotely. The Workstation clients allow calling services within a transaction by having the full power of Tuxedo at your fingertips. It has the same API we used to develop native clients and servers. The only limitation is that it requires Tuxedo libraries installed to work. We also created a web service running as a Tuxedo server in a few lines of Python code. While sacrificing the power of Tuxedo, we gained the interoperability of REST services. Now you are equipped with several options to expose Tuxedo services and ready for future challenges.

In the next chapter, we will learn how to consume web services in a Tuxedo application. While it is simple to do in Python, there are some tricks to make them fit better with the rest of the Tuxedo application.

Questions

1. How does the native Tuxedo client know to connect to the application?

2. How does the Workstation client know to connect to the application?

3. What is the name of the process the Workstation client interacts with?

Further reading

- Workstation client documentation: `https://docs.oracle.com/cd/E53645_01/tuxedo/docs12cr2/ws/wsapp.html`

- WSL server configuration: `https://docs.oracle.com/cd/E53645_01/tuxedo/docs12cr2/rf5/rf5.html`

11
Consuming External Services in Tuxedo

A Tuxedo application does not live in isolation; it has to interact with other parts of the system and external systems. In *Chapter 10*, *Accessing the Tuxedo Application*, we learned how to consume services provided by Tuxedo from other applications. Now we will change the direction of calls and consume external services in our Tuxedo application. Of all chapters in the book, this has the least to do with Tuxedo, having more to do with writing Python code itself. But practice makes perfect, and we will develop a couple of applications.

In this chapter, we will cover the following topics:

- Consuming services
- Handling stateful protocols

By the end of this chapter, you will know how to structure your code to best fit a Tuxedo application. You will also learn how you can use Tuxedo message queues to add non-functional features. Finally, you will learn how to use multi-threaded servers to keep the state for connections.

Technical requirements

All code for this chapter is available at `https://github.com/PacktPublishing/Modernizing-Oracle-Tuxedo-Applications-with-Python/tree/main/Chapter11`.

The Code in Action video for the chapter can be found at `https://bit.ly/3vzK2kd`.

Consuming services

On the surface, consuming services using Python is easy. There are plenty of libraries available for all protocols and data formats. Consuming external services or interacting with internal microservices is just a part of everyday work these days. As you try to do the same in Tuxedo, you might feel tempted to call one or more REST or SOAP services directly from your application code. The service will work just fine, but you will lose features that make Tuxedo applications great.

The best way to consume external services is by creating proxy services for them in Tuxedo and hiding the external call behind a familiar Tuxedo service call. It makes application maintenance easier. As you know from *Chapter 7, Distributed Transactions*, most services are subject to two timeouts – the *blocking timeout* and the *transaction timeout*. Proxy services should not participate in the transaction and the transaction timeout will not apply. However, the blocking timeout can still happen while waiting for an external service to respond. It is important to specify a timeout for an external service that is shorter than the blocking timeout. And it is much easier to do with proxy services when all they do is just call the external service.

There is another useful feature of proxy services: you can use them to replace existing Tuxedo services with an external implementation in a different language and technology stack. And thanks to the proxy service, the rest of the application will not have to be adjusted.

We will start with a simple application that offers a proxy service for currency conversion rates offered by **European Central Bank** (**ECB**) at `https://www.ecb.europa.eu/stats/eurofxref/eurofxref-daily.xml`. Unlike many other services, this is just simple XML content and can be accessed by everyone without signing up for the service.

Like every Tuxedo application, it needs a configuration. The application configuration in
ubbconfig is quite simple with a single group and server:

```
*RESOURCES
MASTER tuxapp
MODEL SHM
IPCKEY 32769
*MACHINES
"15c365dcb562" LMID=tuxapp
    TUXCONFIG="/home/oracle/code/11/tuxconfig"
    TUXDIR="/home/oracle/tuxhome/tuxedo12.2.2.0.0"
    APPDIR="/home/oracle/code/11"
*GROUPS
GROUP1 LMID=tuxapp GRPNO=1 TMSNAME=TMS TMSCOUNT=2
*SERVERS
"ecb.py" SRVGRP=GROUP1 SRVID=1
    REPLYQ=Y MAXGEN=2 RESTART=Y GRACE=0
    MIN=1 MAX=1
```

With the configuration complete, we need a service to retrieve the data. The GETRATES
service retrieves the currency rates provided by the European Central Bank, parses the
XML data, and stores it in Tuxedo FML32 fields. For the simplicity of the example, data is
stored in the TA_STATE field. Here is the code for the ecb.py server:

```python
#!/usr/bin/env python3
import sys
import urllib.request
from xml.etree import ElementTree as et
import tuxedo as t
class Server:
    def tpsvrinit(self, args):
        t.tpadvertise("GETRATES")
        return 0
    def GETRATES(self, data):
        try:
            f = urllib.request.urlopen(
                "https://www.ecb.europa.eu"
                + "/stats/eurofxref/eurofxref-daily.xml",
```

```
            timeout=10,
        )
        rates = et.fromstring(f.read().decode("utf8"))
        data["TA_STATE"] = ""
        for r in rates.findall(".//*[@currency]"):
            data["TA_STATE"] += "{}={};".format(
                r.attrib["currency"], r.attrib["rate"]
            )
    except:
        return t.tpreturn(t.TPFAIL, 0, {})
    else:
        return t.tpreturn(t.TPSUCCESS, 0, data)
if __name__ == "__main__":
    t.run(Server(), sys.argv)
```

The most important parts of the code, however, are the following:

- A 10-second timeout for the ECB data retrieval operation. It must be smaller than the application blocking timeout. That prevents situations when the Tuxedo application has timed out and does not expect a response, but the service is still waiting for external service to respond.

- Depending on the success or exception case, the service returns a TPSUCCESS or TPFAIL code. This allows Tuxedo to be aware of errors and better integrates with other parts of the application.

Once you have started the application, we can proceed to using it. Let us call the GETRATES service and see if it works as expected:

```
import tuxedo as t
print (t.tpcall("GETRATES", {}).data)
```

If everything goes well, you will see a long list of currencies and their conversion rates, as in, *USD=1.2121;JPY=128.83;BGN=1.9558;...* But what if it does not go well?

Adding fault tolerance for external services

External services are outside our control and the network path is long, with plenty of points of failure. Since we wrapped the external service in a Tuxedo proxy service, we can utilize the Tuxedo message queue we learned about in *Chapter 8, Using Tuxedo Message Queue*, to achieve some fault tolerance.

We will use the store and forward capability to retry multiple currency rate downloads multiple times until it succeeds. Let's create a queue space and queues with the following commands:

```
export QMCONFIG=`pwd`/qmconfig
echo "crdl $QMCONFIG 0 200" | qmadmin
echo "qspc QSPACE 239999 100 3 5 5 100 ERR y 16" | qmadmin
echo -e "qopen QSPACE\nqcr ERR time none 0 0 100% 0% ''" |
qmadmin
echo -e "qopen QSPACE\nqcr GETRATES time none 100 5 100% 0% ''"
| qmadmin
echo -e "qopen QSPACE\nqcr GETRATES_DONE time none 3 5 100% 0%
''" | qmadmin
```

The GETRATES queue will be used to call the GETRATES service up to 100 times with a 5-second interval. If it does not succeed after 100 retries, the message will be put into the ERR queue and the application administrator has to keep an eye on it. The third queue, called GETRATES_DONE, will be used to decouple the GETRATES service from the consumer of the currency rates, so we can use a different one, or several consumers, in the future. We will ask the Tuxedo /Q component to put the GETRATES response into this queue without any changes to the service itself.

To use the Tuxedo message queue, we have to create a new application configuration file, ubbconfig, with the following content:

```
*RESOURCES
MASTER tuxapp
MODEL SHM
IPCKEY 32769
*MACHINES
"15c365dcb562" LMID=tuxapp
    TUXCONFIG="/home/oracle/code/11/tuxconfig"
    TUXDIR="/home/oracle/tuxhome/tuxedo12.2.2.0.0"
    APPDIR="/home/oracle/code/11"
    TLOGDEVICE="/home/oracle/code/11/tlog"
*GROUPS
GROUP1 LMID=tuxapp GRPNO=1 TMSNAME=TMS TMSCOUNT=2
GROUP2 LMID=tuxapp GRPNO=2 TMSNAME=TMS_QM TMSCOUNT=2
    OPENINFO="TUXEDO/QM:/home/oracle/code/11/qmconfig:QSPACE"
*SERVERS
```

```
"ecb.py" SRVGRP=GROUP1 SRVID=1
    REPLYQ=Y MAXGEN=2 RESTART=Y GRACE=0
"app.py" SRVGRP=GROUP1 SRVID=2
    REPLYQ=Y MAXGEN=2 RESTART=Y GRACE=0
TMQUEUE SRVGRP=GROUP2 SRVID=3 REPLYQ=Y MAXGEN=2
    RESTART=Y GRACE=0 CLOPT="-s QSPACE:TMQUEUE -- "
TMQFORWARD SRVGRP=GROUP2 SRVID=4
    REPLYQ=N MAXGEN=2 RESTART=Y GRACE=0
    CLOPT="-A -- -q GETRATES,GETRATES_DONE -i 1"
```

From the preceding code, we see that in addition to the ecb.py server, we added another one called app.py that will provide the GETRATES_DONE service. And we also include Tuxedo configuration for queues and a queue-forwarding servers that reads messages from two queues and sends them to the services with the same name. One queue is for the GETRATES service and the second for the GETRATES_DONE service.

For queues to work, we need a transaction log created with the following commands:

```
tmloadcf -y ubbconfig
echo crdl -z `pwd`/tlog -b 200 | tmadmin
echo crlog -m tuxapp | tmadmin
```

The final piece missing is the app.py server with the following code:

```
#!/usr/bin/env python3
import sys
import tuxedo as t
class Server:
    def tpsvrinit(self, args):
        t.tpadvertise("GETRATES_ASYNC")
        t.tpadvertise("GETRATES_DONE")
        return 0
    def GETRATES_ASYNC(self, data):
        t.tpenqueue("QSPACE", "GETRATES",
            t.TPQCTL(
                replyqueue="GETRATES_DONE",
                flags=t.TPQREPLYQ,
            ),
            data,
```

```
        )
        return t.tpreturn(t.TPSUCCESS, 0, {})
    def GETRATES_DONE(self, data):
        t.userlog(data["TA_STATE"][0])
        return t.tpreturn(t.TPSUCCESS, 0, {})
            if __name__ == "__main__":
    t.run(Server(), sys.argv)
```

The GETRATES_DONE service will be called once the GETRATES service succeeds and receives the currency conversion rates. We will just write them in ULOG. The second service GETRATES_ASYNC is a proxy service for calling the GETRATES proxy service. To paraphrase Butler Lampson, *"All problems in the Tuxedo application can be solved by another level of proxy services"*. Here the GETRATES_ASYNC proxy service hides the queue space, queue, reply queue, and other details needed for the fault-tolerant version of the external service call.

Once you start the application, you can schedule the fault-tolerant ECB currency conversion rate download. Instead of retrieving rates directly, we place the request into a queue:

```
import tuxedo as t
t.tpcall("GETRATES_ASYNC", {})
```

After that, Tuxedo queue forwarding starts to work and calls the GETRATES service. Once it succeeds, the GETRATES_DONE service will be called. In a few seconds, ULOG will contain the latest data from ECB.

By wrapping the external service in a way that is native to Tuxedo, we still have the option to call the GETRATES service and receive the result immediately. Or, we use the Tuxedo queuing component to add fault tolerance with no changes to the GETRATES service.

Most web service protocols are based on a single request and response and can be implemented with a stateless single-threaded Tuxedo server. But there are several that have a complicated protocol or a stateful session and require the use of multiple threads. In the next section, we'll learn about them.

Handling stateful protocols

The HTTP protocol supports persistent connections that reuse the same underlying TCP/IP connection for multiple request/response pairs. Sometimes you need to persist session cookies across requests. There are still a lot of network protocols that use a persistent TCP/IP connection for multiple asynchronous requests and responses. One way or another, there are times when you need to build a multi-threaded Tuxedo server.

To keep our examples simple, we will use a simple TCP/IP connection and for each character the server receives, it will reply with the same uppercase character. It will be our TCP/IP TOUPPER service. Here is the code for toupper.py, which listens on port 8765:

```
import socket
sock = socket.socket()
sock.bind(("", 8765))
sock.listen(1)
while True:
    con, _ = sock.accept()
    while True:
        c = con.recv(1024)
        r = c.decode().upper().encode()
        con.sendall(r)
```

You can start this server in the background by executing the following command:

```
python3 toupper.py &
```

Now we can create a client for this service in the out.py Tuxedo server:

```
#!/usr/bin/env python3
import socket, sys, threading, time
import tuxedo as t
class Server:
    def tpsvrinit(self, args):
        t.tpadvertise("TOUPPER")
        self.lock = threading.Lock()
        self.sock = socket.create_connection((
                    "127.0.0.1", 8765))
        threading.Thread(target=self.ping,
```

```
                    daemon=True).start()
        return 0
    def ping(self):
        while True:
            time.sleep(60)
            with self.lock:
                self.sock.sendall(b"ping")
    def TOUPPER(self, data):
        req = data.encode()
        with self.lock:
            self.sock.sendall(req)
            res = self.sock.recv(len(req))
        return t.tpreturn(t.TPSUCCESS, 0, res.decode())
if __name__ == "__main__":
    t.run(Server(), sys.argv)
```

To keep it simple, we have skipped all error-handling code. The server will hold a single TCP/IP connection to the destination service. Since the server will be multi-threaded, we will need a `threading.Lock` object to serialize access to the connection.

Protocols with persistent TCP/IP connections usually have periodic heartbeat messages to ensure the idle connection has not been dropped and to reconnect if it has been dropped. In our case, we will send a `ping` message every `60` seconds. We have also a `TOUPPER` service as a proxy for the external service. To avoid mixing requests and responses from multiple threads, we serialize access to the external service by taking a lock.

A stateful client for a real protocol would be more complicated, but it should be obvious that Tuxedo does not stand in your way and you can develop your code just as you would in a non-Tuxedo application.

We will also need an `ubbconfig` configuration file to start the `out.py` server:

```
*RESOURCES
MASTER tuxapp
MODEL SHM
IPCKEY 32769
*MACHINES
"15c365dcb562" LMID=tuxapp
    TUXCONFIG="/home/oracle/code/11/tuxconfig"
    TUXDIR="/home/oracle/tuxhome/tuxedo12.2.2.0.0"
    APPDIR="/home/oracle/code/11"
*GROUPS
GROUP1 LMID=tuxapp GRPNO=1 TMSNAME=TMS TMSCOUNT=2
*SERVERS
"out.py" SRVGRP=GROUP1 SRVID=1
    REPLYQ=Y MAXGEN=2 RESTART=Y GRACE=0
    MIN=1 MAX=1
    MINDISPATCHTHREADS=2 MAXDISPATCHTHREADS=2
```

Once you have started the application, we can test it as follows:

```
import tuxedo as t
print (t.tpcall("TOUPPER", "Hello TCP/IP world!").data)
```

You will get a `HELLO TCP/IP WORLD!` message in response. The code works as it should, but we skipped a few things besides the error handling that would make the `TOUPPER` proxy service a good citizen of the Tuxedo application: it should use a timeout for calling external services and it should return `TPFAIL` in the case of an error.

Summary

In this chapter, we learned how to wrap external services in Tuxedo service proxies to provide a familiar Tuxedo interface. We learned to pay attention to explicitly setting timeouts and returning a correct `TPFAIL` code in case of error. We also learned to use Tuxedo messages queues for fault-tolerance and asynchronous processing. And finally, we created a stateful multi-threaded server to take advantage of persistent connections. Now you should be equipped with the knowledge to access external services using any protocol supported by Python libraries, which is a lot more than C or COBOL programming languages offer.

In the next chapter, we will combine what we learned in *Chapter 10, Accessing the Tuxedo Application,* and this chapter to create a bi-directional gateway between Tuxedo and the NATS.io messaging system.

Questions

1. Which pattern was used to wrap external services?

2. Which out of *blocking* and *transaction* timeouts are more likely to happen?

3. What code must the `tpreturn` function return for a service call to be retried by the `TPQFORWARD` server?

12
Modernizing the Tuxedo Applications

Oracle Tuxedo is an interesting piece of software. It has benefited from years of research and development, it works in synergy with Unix-like operating systems, and it was efficient on computers less powerful than today's smartphones. I like it a lot. But it is commercial and closed source software. That alone prevents it from being used when a new software is developed.

Most new software projects rely on open source software with an active community. Sooner or later, someone from management will initiate the *modernization* of Tuxedo applications, which stands for *replacing with cheaper alternatives* in management-speak.

In this chapter, we will attempt to replace parts of a Tuxedo application by covering the following topics:

- Introducing NATS
- Developing a basic NATS application
- Developing a bi-directional gateway

By the end of this chapter, you will know about NATS and why it is a good fit for modernizing Tuxedo applications. You will learn how to develop a NATS application and how it compares to developing a Tuxedo application. And finally, you will create and use a bi-directional gateway that allows you to call Tuxedo from NATS and the other way around.

Technical requirements

All code for this chapter is available at `https://github.com/PacktPublishing/Modernizing-Oracle-Tuxedo-Applications-with-Python/tree/main/Chapter12`.

The Code in Action video for the chapter can be found at `https://bit.ly/3czuBzC`.

Introducing NATS

NATS is an open source modern messaging system. It allows messages to be sent and received between different parts of the system. It is the backbone of several microservice frameworks and is used by many organizations to develop distributed systems. NATS is a **Cloud Native Computing Foundation** project along with Kubernetes, Prometheus, and other popular names. It is lightweight, small in size, and high performance. NATS applications can be developed in dozens of programming languages with libraries maintained by the NATS project itself or the community. As long as you can create a TCP/IP connection and send and receive text over it, you can use NATS. You don't need to link your code with the vendor's libraries.

Just like Tuxedo, NATS has at-most-once delivery semantics with no message persistence. It also supports request-response or request-reply semantics, just like Tuxedo does with the `tpcall` function.

Unlike Tuxedo, it does not support transactions, and that is the greatest loss when moving an application from Tuxedo to NATS. It also sends and receives messages over the TCP/IP protocol and has a message broker where Tuxedo has none, and uses System V IPC queues on the same physical machine. The only payload type supported is raw bytes, and there is nothing similar to typed buffers in Tuxedo, but that is not a problem at all. Using JSON converted to bytes is enough for most needs.

Because NATS is a modern messaging system, you can get it up and running with the following single command:

```
docker run -d --name nats-main -p 4222:4222 nats
```

This will download and start the latest version of the NATS server listening on port 4222. To use NATS from Python, we will need a library. Several libraries are available, but we will use the one provided and supported by the NATS project, even though it uses a more complicated implementation with coroutines. You can install it using the following command:

```
sudo pip3 install asyncio-nats-client
```

So, we see that, in only two simple steps, we are ready to develop a NATS application. Let's see how we actually develop it now.

Developing a basic NATS application

We started learning Tuxedo by developing a TOUPPER service first. We will do the same for NATS to compare where it is similar to Tuxedo and see what the main differences are.

Let's create a toupper.py application with the following content:

```python
import asyncio
from nats.aio.client import Client
nc = Client()


async def TOUPPER(msg):
    await nc.publish(msg.reply, msg.data.decode().upper().
encode())


async def run(loop):
    await nc.connect(
        servers=["nats://host.docker.internal:4222"], loop=loop
    )
    await nc.subscribe("TOUPPER", cb=TOUPPER)
    try:
        msg = await nc.request(
            "TOUPPER",
            "Hello NATS!".encode(),
            timeout=5,
        )
        print(msg.data.decode())
```

```
    except asyncio.TimeoutError:
        print("Timed out waiting for response")
    await nc.close()


loop = asyncio.get_event_loop()
loop.run_until_complete(run(loop))
loop.close()
```

The TOUPPER function is what handles the incoming requests. It ends by sending a response to the reply topic. This is what the tpreturn function does in Tuxedo, although Tuxedo returns more information along with the message. The message payload supports only bytes and we receive the input string as bytes. We have to create a string first and convert it into an uppercase form. To return the result, we have to convert the string back to bytes.

In the run function, we connect our application to the NATS message broker that should be running on Docker network port 4222. Tuxedo is a broker-less messaging system and this step was not needed.

Next, we subscribe to the TOUPPER topic by calling the subscribe function and we use the TOUPPER function to handle requests. In Tuxedo, we advertised services using the tpadvertise function with the same name as the handler method. There are no strict requirements for using the same name for the topic and the function name, but we do it anyway out of convenience.

To call the TOUPPER service, we send a request using the NATS request function and pass the message itself as bytes in the second argument. We used the tpcall function in Tuxedo to call services. We can also specify a timeout for each request function call, similarly to blocking timeouts in Tuxedo.

The rest of the code uses the asyncio library to run coroutines, and it is not related to NATS itself but the Python library we chose to interact with NATS.

We do not need a configuration file or separate files for the client and server. Running this Python code will both start the server part and run the client that calls the service:

```
python3 toupper.py
```

You will get a HELLO NATS.IO! output, proving that your code works. The first application was quite easy. But integrating it with Tuxedo will be more complicated.

Developing a bi-directional gateway

NATS sends a bunch of bytes around, but Tuxedo requests and responses carry around more information. For both systems to communicate seamlessly, we have to advance NATS messages:

- Requests in Tuxedo have flags such as TPTRAN and TPNOREPLY that indicate that the service call is a part of a global transaction and that no response is expected from the service. NATS does not support transactions, so we do not need the TPTRAN flag, but TPNOREPLY is useful.

- Tuxedo supports multiple typed buffers. We could find a way to encode CARRAY, STRING, and FML32 messages but for simplicity, we will support only FML32 messages that are sent as Python dictionaries.

- Responses in Tuxedo have the rval, rcode, and data fields, and we will need them all in our gateway.

To include all extra information in NATS messages, we will use JSON data converted to bytes. The service request will have the following format:

```
{"flags":"TPNOREPLY", "data":{}}
```

And the response will have the following format:

```
{"rval":"TPSUCCESS", "rcode":0, "data":{}}
```

The future version 2.2 of NATS supports message headers, and we could put all extra fields there, leaving the message payload for the message itself. But that version is not released at the time that I am writing this book, so we will have to put all fields in the message payload itself.

To call a service implemented using NATS, we will use the proxy pattern we learned about in *Chapter 11, Consuming External Services in Tuxedo*. A Tuxedo service will send a message to a NATS topic with the same name and wait for the response from NATS before returning it to the caller.

To call a Tuxedo service, we will create a thread that listens on a NATS topic. Once a message is received from NATS, it will call the service with the same name as the NATS topic and send the service response back to the reply topic back to NATS. This is just like what we did with an HTTP server in *Chapter 10, Accessing the Tuxedo Application*.

Our application, which combines Tuxedo and NATS into one, will have two TUX1 and TUX2 services implemented on the Tuxedo side and two NATS1 and NATS2 services (topics) implemented on the NATS side.

Let's create a gateway server first that will make this interaction seamless. The code is long and looks scary, so we will break it down one function at a time. It will be a Tuxedo server called nat sgw.py with the following content:

```python
#!/usr/bin/env python3
import asyncio
import json
import sys
import threading
import tuxedo as t
from nats.aio.client import Client

nc = Client()
loop = asyncio.get_event_loop()

async def on_message(msg):
    req = json.loads(msg.data.decode())
    svcname = msg.subject
    data = req["data"]
    try:
        rval, rcode, data = await loop.run_in_executor(
            None, t.tpcall, svcname, data
        )
    except t.XatmiExceptio:
        rval, rcode, data = t.TPESVCFAIL, 0, None
    await nc.publish(
        msg.reply,
        json.dumps(
            {
                "rval": "TPFAIL"
                if rval == t.TPESVCFAIL
                else "TPSUCCESS",
                "rcode": rcode,
                "data": data,
            }
        ).encode(),
    )
```

The on_message function is called for each message received from NATS. The subject field of the message contains the topic name, and we will use it as the name of the Tuxedo service to be called. We unpack the payload JSON to get the message for calling the service. The actual service tpcall is performed from the run_in_executor function to safely perform a blocking function call from the asyncio event loop. Once a response or an error is received, the response for NATS is packed in JSON, converted to bytes, and published on the reply topic. Then the on_service function is called when a service call is made from Tuxedo to send over to NATS:

```
async def on_service(name, data, flags):
    if flags & t.TPTRAN:
        return (t.TPFAIL, 0, data)
    try:
        msg = await nc.request(
            name,
            json.dumps({"flags": "", "data": data}).encode(),
            timeout=5,
        )
        res = json.loads(msg.data.decode())
        return (
            t.TPSUCCESS if res["rval"] == "TPSUCCESS" else
t.TPFAIL,
            res["rcode"],
            res["data"],
        )
    except asyncio.TimeoutError:
        return (t.TPFAIL, 0, data)
```

First, it safeguards against calls made in a transaction. A request is sent to NATS with the Tuxedo message packed in JSON and converted to bytes. Once a response is received from NATS, the response fields are extracted from JSON and returned to the caller:

```
class Server:
    def tpsvrinit(self, args):
        threading.Thread(
            target=loop.run_forever, daemon=True
        ).start()
        asyncio.run_coroutine_threadsafe(
```

```
        nc.connect(
            servers=["nats://host.docker.internal:4222"],
            loop=loop,
        ),
        loop,
    ).result()

    setattr(self, "NATS1", self.PROXY)
    t.tpadvertise("NATS1")
    setattr(self, "NATS2", self.PROXY)
    t.tpadvertise("NATS2")

    asyncio.run_coroutine_threadsafe(
        nc.subscribe("TUX1", cb=on_message), loop
    ).result()
    asyncio.run_coroutine_threadsafe(
        nc.subscribe("TUX2", cb=on_message), loop
    ).result()
    return 0
```

Then, from the preceding code, we see that during the server initialization in the tpsvrinit method, a connection to NATS is established. Two services named NATS1 and NATS2 are advertised. These proxy services will forward service requests to NATS. Then, we subscribe to the TUX1 and TUX2 topics in NATS. The server will call Tuxedo services with these names once new messages are received on those topics:

```
def PROXY(self, data, name, flags):
    future = asyncio.run_coroutine_threadsafe(
        on_service(name, data, flags), loop
    )
    rval, rcode, data = future.result(timeout=5)
    return t.tpreturn(rval, rcode, data)

if __name__ == "__main__":
    t.run(Server(), sys.argv)
```

The PROXY method is just a fancy way of calling the on_service function in the asyncio event loop from a Tuxedo thread. Again, this has nothing to do with Tuxedo or NATS, but rather the NATS library we chose.

The gateway itself is complete now. To test it, we will implement the TUX1 and TUX2 services in Tuxedo, which can be called from NATS by using the gateway. We will put them inside the app.py Tuxedo server with the following content:

```python
#!/usr/bin/env python3
import sys
import tuxedo as t
class Server:
    def tpsvrinit(self, args):
        t.tpadvertise("TUX1")
        t.tpadvertise("TUX2")
        return 0
    def TUX1(self, data):
        if "TA_STATUS" not in data:
            data["TA_STATUS"] = []
        data["TA_STATUS"].append("Hello from TUX1")
        return t.tpreturn(t.TPSUCCESS, 0, data)
    def TUX2(self, data):
        data["TA_STATUS"] = ["Hello from TUX2"]
        _, _, data = t.tpcall("NATS1", data, t.TPNOTRAN)
        return t.tpreturn(t.TPSUCCESS, 0, data)
if __name__ == "__main__":
    t.run(Server(), sys.argv)
```

The TUX1 service returns a greeting message to the caller. The TUX2 service does the same but calls the NATS1 service to get a greeting message from it before returning to the caller. It does not look complicated so far.

But now we need the NATS1 service that is called from the TUX2 service, and we will implement it in NATS without using any of the Tuxedo libraries. So create a natsapp.py file with the following content:

```python
import asyncio
import json
from nats.aio.client import Client
```

```
nc = Client()

async def NATS1(msg):
    data = json.loads(msg.data.decode())["data"]
    if "TA_STATUS" not in data:
        data["TA_STATUS"] = []
    data["TA_STATUS"].append("Hello from NATS1")
    await nc.publish(
        msg.reply,
        json.dumps(
            {"rval": "TPSUCCESS", "rcode": 0, "data": data}
        ).encode(),
    )
```

The basic NATS application structure should be familiar now after building the
`toupper.py` application. The NATS1 service adds a greeting to the caller. The main
complexity comes from packing and unpacking the additional fields needed by Tuxedo
into JSON and converting it to and from bytes. For completeness, we also have the NATS2
service, which adds a greeting:

```
async def NATS2(msg):
    data = json.loads(msg.data.decode())["data"]
    data["TA_STATUS"] = ["Hello from NATS2"]
    msg2 = await nc.request(
        "TUX1",
        json.dumps({"flags": "", "data": data}).encode(),
        timeout=5,
    )
    data = json.loads(msg2.data.decode())["data"]

    await nc.publish(
        msg.reply,
        json.dumps(
            {"rval": "TPSUCCESS", "rcode": 0, "data": data}
        ).encode(),
    )
```

And then `NATS2` calls the `TUX1` Tuxedo service to get a greeting from it before returning the response. Here again, most of the complexity comes from packing data into JSON. Next is the connection:

```
async def init(loop):
    await nc.connect(
        servers=["nats://host.docker.internal:4222"], loop=loop
    )
    await nc.subscribe("NATS1", cb=NATS1)
    await nc.subscribe("NATS2", cb=NATS2)
loop = asyncio.get_event_loop()
loop.run_until_complete(init(loop))
loop.run_forever()
```

And we complete the NATS application by connecting to the NATS server, subscribing to the `NATS1` and `NATS2` topics, and running the `asyncio` event loop.

Finally, we can create the `ubbconfig` file for Tuxedo:

```
*RESOURCES
MASTER tuxapp
MODEL SHM
IPCKEY 32769
*MACHINES
"15c365dcb562" LMID=tuxapp
    TUXCONFIG="/home/oracle/code/12/tuxconfig"
    TUXDIR="/home/oracle/tuxhome/tuxedo12.2.2.0.0"
    APPDIR="/home/oracle/code/12"
*GROUPS
GROUP1 LMID=tuxapp GRPNO=1 TMSNAME=TMS TMSCOUNT=2
*SERVERS
"app.py" SRVGRP=GROUP1 SRVID=1
    REPLYQ=Y MAXGEN=2 RESTART=Y GRACE=0
    MIN=1 MAX=1
"natsgw.py" SRVGRP=GROUP1 SRVID=10
    REPLYQ=Y MAXGEN=2 RESTART=Y GRACE=0
    MIN=1 MAX=1
    MINDISPATCHTHREADS=2 MAXDISPATCHTHREADS=2
```

Once you have the configuration, you can start the Tuxedo application. This time you will have to start the NATS application. There are better ways to run it and ensure it gets restarted after failures, but for now, we will use the following command:

```
python3 natsapp.py&
```

Now we have a running NATS messaging system, a running Tuxedo application, and a running NATS application. We can start testing to see how it works:

```
import tuxedo as t
t.tpcall("TUX1", {}).data
```

You will get a greeting from the TUX1 service running in Tuxedo. Nothing interesting so far. Let's call the next service:

```
t.tpcall("NATS1", {}).data
```

This will show you a greeting from the NATS1 service. The request traveled to the NATS1 Tuxedo service that acts as a proxy, publishing a message to a topic in the NATS messaging system. The NATS application received it from the messaging system, added a greeting, and sent it back to us. We know that we can call services implemented in NATS from Tuxedo clients.

Let's try a more complicated example using the following code:

```
t.tpcall("TUX2", {}).data
```

It will return a greeting from the TUX2 service and the NATS1 service. The request was first received in the TUX2 service running Tuxedo and then sent to the NATS1 proxy service. And then it traveled to the NATS application over the NATS messaging system. Calling the NATS application from the Tuxedo server works fine as well.

For the final example, let's call the NATS2 service:

```
t.tpcall("NATS2", {}).data
```

You will get a greeting from the NATS2 service first and then from the TUX1 service. The request traveled from the Tuxedo application to the NATS application, and then from the NATS application to the Tuxedo application. And then the response messages traveled from the Tuxedo application back to the NATS application, then back to the Tuxedo application. If your head is starting to spin, here is a diagram that shows what just happened:

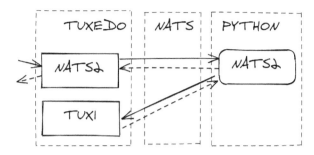

Figure 12.1 – The NATS2 service call

At this point, we have successfully called a NATS service from Tuxedo and a Tuxedo service from NATS. Our service call chain went back and forth over the NATS messaging system. The Tuxedo services we used looked like regular Tuxedo services. The NATS services had more work to do because Tuxedo has more request and response fields and there are no shortcuts to use them in NATS.

We managed to integrate Tuxedo and NATS applications through a gateway in less than 100 lines of Python code. By that, we achieved the goal of the chapter. But imagine what could be done with 100 more lines of code!

Summary

In this chapter, we learned about the NATS messaging system, which looks modern and promising today. Just like Tuxedo did decades ago. We learned about some similarities and major differences and how developing a NATS application compares to developing a Tuxedo application. Many areas have to be addressed for the NATS application, such as starting and stopping it, monitoring it, and configuring it. But this book is not about NATS. We learned how to integrate NATS and Tuxedo applications to call each other.

After this chapter, you are given a choice of how to implement new features in existing applications: either on the Tuxedo side or the NATS side. Even if you choose to use a different messaging system, you have knowledge and directions on how to start working on this task. And most importantly, using Python makes the task reasonably easy and painless. Take it from someone who has done it in C++.

And with that, we have reached the end of this book. There are more standard and paid add-on features but they are all built around the same core features that we learned about and explored in this book. Many of them look scary because of the large amount of unforgiving C program code required to use the functionality. But with Python, the sky's the limit!

Questions

1. What is the NATS analog of the Tuxedo `tpadvertise` function?
2. What is the NATS analog of the Tuxedo `tpcall` function?
3. What is the NATS function to perform what the Tuxedo `tpreturn` function does?

Further reading

- Comparing NATS: `https://docs.nats.io/compare-nats`
- NATS Docker image: `https://hub.docker.com/_/nats`

Assessments

Chapter 1

1. C, COBOL, C++, Java, Python, PHP, Ruby

2. More than 30 years

3. System V IPC queues

Chapter 2

1. `TUXCONFIG`

2. `tmloadcf`

3. `tmboot`

4. `tmshutdown`

5. `ps aux | grep ULOG`

Chapter 3

1. SIGTERM

2. No

3. MSSQ

4. SSSQ

Chapter 4

1. `STRING, CARRAY, and FML32`

2. `ud32`

3. `tpimport`

4. `Ffloatev32`

Chapter 5

1. `tpadvertisex`
2. No
3. `TPNOREPLY`
4. `TPGETANY`
5. `TPNOBLOCK`

Chapter 6

1. `tpadmcall`
2. `tpcall`
3. Size of result
4. `MIB_LOCAL`

Chapter 7

1. `TPNOTRAN`
2. `TPNOTIME`
3. `SCANUNIT`
4. No

Chapter 8

1. `tpenqueue`
2. `tpenqueue`
3. Correlation identifier
4. `TPQWAIT`

Chapter 9

1. `buildtms`
2. `dba_pending_transactions`
3. Use *****

Chapter 10

1. `TUXCONFIG` environment variable
2. `WSNADDR` environment variable
3. `WSH`

Chapter 11

1. Proxy
2. Blocking timeout
3. `TPFAIL`

Chapter 12

1. `subscribe`
2. `request`
3. `publish`

`Packt.com`

Subscribe to our online digital library for full access to over 7,000 books and videos, as well as industry leading tools to help you plan your personal development and advance your career. For more information, please visit our website.

Why subscribe?

- Spend less time learning and more time coding with practical eBooks and Videos from over 4,000 industry professionals

- Improve your learning with Skill Plans built especially for you

- Get a free eBook or video every month

- Fully searchable for easy access to vital information

- Copy and paste, print, and bookmark content

Did you know that Packt offers eBook versions of every book published, with PDF and ePub files available? You can upgrade to the eBook version at `packt.com` and as a print book customer, you are entitled to a discount on the eBook copy. Get in touch with us at `customercare@packtpub.com` for more details.

At `www.packt.com`, you can also read a collection of free technical articles, sign up for a range of free newsletters, and receive exclusive discounts and offers on Packt books and eBooks.

Other Books You May Enjoy

If you enjoyed this book, you may be interested in these other books by Packt:

Modern Python Cookbook – Second Edition

Steven F. Lott

ISBN: 978-1-80020-745-5

- See the intricate details of the Python syntax and how to use it to your advantage
- Improve your coding with Python readability through functions
- Manipulate data effectively using built-in data structures
- Get acquainted with advanced programming techniques in Python
- Equip yourself with functional and statistical programming features
- Write proper tests to be sure a program works as advertised
- Integrate application software using Python

Oracle CX Cloud Suite

Kresimir Juric

ISBN: 978-1-78883-493-3

- Differentiate between Oracle CRM and CX Cloud suites

- Explore a variety of Oracle CX Cloud tools for marketing and sales

- Set up users and database connections during deployment

- Employ Cloud Suite CX tools to aid in planning and analysis

- Implement hybrid Oracle CX solutions and connect with legacy systems

- Integrate with social media connectors like Facebook and LinkedIn

- Leverage Oracle ICS and Oracle CX Suite to improve business results

Packt is searching for authors like you

If you're interested in becoming an author for Packt, please visit `authors.packtpub.com` and apply today. We have worked with thousands of developers and tech professionals, just like you, to help them share their insight with the global tech community. You can make a general application, apply for a specific hot topic that we are recruiting an author for, or submit your own idea.

Leave a review - let other readers know what you think

Please share your thoughts on this book with others by leaving a review on the site that you bought it from. If you purchased the book from Amazon, please leave us an honest review on this book's Amazon page. This is vital so that other potential readers can see and use your unbiased opinion to make purchasing decisions, we can understand what our customers think about our products, and our authors can see your feedback on the title that they have worked with Packt to create. It will only take a few minutes of your time, but is valuable to other potential customers, our authors, and Packt. Thank you!

Index

Q

R

S

T